SpringerBriefs in Plant Science

More information about this series at http://www.springer.com/series/10080

Bethany Gwen Elkington • Djaja Djendoel Soejarto
Kongmany Sydara

Ethnobotany
of Tuberculosis in Laos

 Springer

Bethany Gwen Elkington
Department of Medicinal Chemistry
and Pharmacognosy
University of Illinois at Chicago
Chicago, IL, USA

Integrative Research Center
Science and Education (Botany)
The Field Museum of Natural History
Chicago, IL, USA

Kongmany Sydara
Ministry of Health
Institute of Traditional Medicine
Vientiane Capital, Laos

Djaja Djendoel Soejarto
Department of Medicinal Chemistry
and Pharmacognosy
University of Illinois at Chicago
Chicago, IL, USA

Integrative Research Center
Science and Education (Botany)
The Field Museum of Natural History
Chicago, IL, USA

ISSN 2192-1229　　　　ISSN 2192-1210 (electronic)
ISBN 978-3-319-10655-7　　ISBN 978-3-319-10656-4 (eBook)
DOI 10.1007/978-3-319-10656-4
Springer Cham Heidelberg Dordrecht London New York

Library of Congress Control Number: 2014948206

Preface

Because the majority of the population of Laos lives in rural areas with limited access to biomedical healthcare, traditional herbal remedies are frequently used. Old Buddhist palm leaf manuscripts provide an invaluable record of these treatments through history, many of which are still used by contemporary traditional healers. This research has explored these herbal medicines of Laos to treat symptoms of tuberculosis, a disease that is currently ravaging the Asian continent. Plant samples were collected for 77 species of plants, which were submitted to various biological assays. This document should help to increase awareness of Laos' rich diversity of medicinal plants and provides incentive for the preservation of the undeveloped forested areas that remain, which still hold a wealth of medical information for future discoveries. This research touched on many aspects of traditional medicine use in Laos and serves as justification and a starting point for further research.

Acknowledgments

First and foremost, we would like to acknowledge the contributions of the participating people of Laos. We have been able to witness firsthand how the use of traditional medicines is a vital part of healthcare in this country. The government of Laos was most kind to grant the necessary permits to conduct interviews, to collect plant samples, and to transport them to the University of Illinois at Chicago (UIC) for analysis.

To the following scientists, we express our thanks for their guidance throughout the entire process of this research. As the director of UIC's Institute of Tuberculosis Research (ITR), Professor Scott Franzblau generously provided necessary resources for the biological assays and much of the structure elucidation portion of this research. Professor Charlotte Gyllenhaal at UIC offered helpful input about ethnobotany, life, health, and sickness. Professor John Hartmann at Northern Illinois University gave important guidance on the linguistics of the palm leaf manuscripts. The SEAsite Laos website (http://www.seasite.niu.edu/lao/), provided through the Center for Southeast Asian Studies at NIU, was a very valuable tool throughout this research. Professor Crystal Patil at UIC was extremely helpful with the anthropology portions of this project. Professor Guido Pauli (UIC) generously provided laboratory space to work in. Professor Hongjie Zhang (UIC; now at Hong Kong Baptist University) kindly gave advice on the chemistry aspects. The NAPRALERT® database, created by the late Professor Norman R. Farnsworth, was extremely helpful. Professor Justin McDaniel (University of Pennsylvania) offered many linguistic and cultural tips, as well as information about the palm leaf manuscripts and Buddhism in Laos. Dr. Mary Riley at UIC was helpful with intellectual property questions and information.

Somsanith Bouamanivong, previously at the Institute of Traditional Medicine (ITM) in Vientiane and now at the Conservation and Botanical Research Center and National Herbarium of Laos, was very helpful in setting up the project. She also helped with taxonomic identification, interviews, and translations. Mr. Bountham Panyachit, previously at the ITM and now at Pharmaceutical Factory No. 2 in Vientiane, proved to be invaluable for his skills in translating Tham-Lao and Tham-Lue, as well as traditional medicine preparation and his ability to find medicinal plants in the wild. The International Union for Conservation of Nature (IUCN) and Dr. Phaivanh Phiapalath paved the way to learn more about plant conservation and the livelihoods of the people in rural Laos through their Livelihoods and Landscapes (LLS) initiative. Mr. Vongtakoune Somsamouth and Dr. Nichole I. Goodsmith also helped tremendously with the LLS field research and many of the medicinal plant photographs featured in this book. The IUCN library in Vientiane also provided important reference resources.

Other experts at the ITM were extremely kind in providing space for our work in Vientiane, in making available their herbarium and numerous literature sources for our use, as well as in generously providing hours and hours of their time and expertise. A huge khop jai, lai lai to the former director of the ITM, Professor Bounhong Southavong, along with Ms. Thongeune Keohavong, Mr. Ounneua Keokongten, Mr. Khamphong Phommavong, Dr. Khamchanh Phonlavong, Mr. Vongtakoune Somsamouth, Mr. Onevilay Souliya, Mr. Khamphanh Thepkaysone, Mr. Manoluck Vanthanouvong, and Dr. Somphao Neunphonsavath at the TMS in Champasak. Sincere thanks to all of the others who have provided help in the process of our research.

We thank the National Library of Laos for allowing us access to and trusting in the use of many Palm Leaf Manuscripts, along with the digital images of the medical manuscripts. Ajan Bounleuth Tammachak and Ajan Thongseui Outhoumphone spent many hours translating entries from the manuscripts into the modern Lao language and script, which were then typed and formatted by Khammack Vongsackda and Khanthamaly Yangnouvong. David Wharton kindly offered help with the digitized manuscript collection. Professor Harald Hundius and Kongdeuane Nettavong also provided invaluable insights into the project.

The Field Museum of Natural History in Chicago allowed access to its herbarium and libraries for use in plant identification, and for curating the plant specimens that were deposited in the John G. Searle Herbarium (F). Dr. James Graham gave helpful advice about many aspects of this research. Darlene Dowdy-Pritchett beautifully mounted all of the plant specimens. We are also indebted to the many plant taxonomists who helped to identify the plant species collected for this research. In particular, the staff at the ITM; Dr. James G. Graham and Dr. Richard H. Ree at the Field Museum Herbarium (F); Dr. Jacinto Regalado at the Missouri Botanical Garden (MBG); Professor Richard M. K. Saunders and Dr. Bine Xue at the University of Hong Kong (HKU); and Professor P.J.A. Kessler at the Nationaal Herbarium Nederlands, Leiden (L).

We thank the following scientists for their help and guidance in our laboratory analysis work. Dr. David Lankin and Dr. Jose Napolitano kindly contributed their NMR expertise. Dr. Chang Hwa Wang always found the time to provide his help and advice on the laboratory portions of this research. Andrew Newsome helped in obtaining and interpreting clear MS results. Dr. Dejan Nikolic was also helpful in the interpretation of MS data. Dr. Charlotte Simmler offered helpful advice on many different aspects of the laboratory research. Bioassays were run by Wei Gao, Marcelle Hon, Yeun Kim, Minjee Kim, and Dennis Pak. We are very thankful to UIC's Research Resources Center (RRC), which allowed us access to state-of-the-art equipment essential for the pharmacognostic analyses.

Many thanks to Eric Stannard, Andy Kwan, and Patti Donofrio at Springer for approaching us with the idea of publishing this research in book form, and for their kind support and persistence in getting to a finished product.

And of course a huge thank you goes to our families, who honor us with their time, encouragement, and support.

Financial support was provided by the International Cooperative Biodiversity Group Grant 2-U01-TW001015-09 awarded to Djaja D. Soejarto as Principal Investigator, as well as the Institute of International Education through a Fulbright Full Grant awarded to Bethany Gwen Elkington, and the National Institutes of Health National Center for Complementary & Alternative Medicine Award Number F31AT006069 awarded to Bethany Gwen Elkington.

While we have received immeasurable help and support from others, the content of this book does not represent the official views of any of the above people or institutions.

Contents

List of Figures

List of Tables

Abbreviations

AD	Anno Domini. Time era also referred to as the "Common Era" (CE)
ADB	Asian Development Bank
BC	Before Christ
BCE	Before the Common Era
CBD	Convention on Biological Diversity
CE	Common Era. Also referred to as AD, beginning in 1 BC in the Julian and Gregorian calendars
CITES	Convention on International Trade in Endangered Species of Wild Fauna and Flora
DMSO	Dimethyl sulfoxide
EtOH	Ethanol
F	Herbarium at the Field Museum of Natural History in Chicago, IL
g	Gram
GMS	Greater Mekong Sub-region
HPLC	High-performance or high-pressure liquid chromatography
ICBG	International Cooperative Biodiversity Group
IPNI	International Plants Name Index
IRB	Institutional Review Board
ISE	International Society of Ethnobiology
ITM	Institute of Traditional Medicine
ITR	Institute for Tuberculosis Research
IUCN	International Union for the Conservation of Nature/The World Conservation Union
L	Liter
Lao PDR	Lao People's Democratic Republic
LC	Liquid chromatography
LLS	Livelihoods and landscapes
LORA	Low-oxygen-recovery assay
LPRP	Lao People's Revolutionary Party
MABA	Microplate Alamar Blue assay
MBSP	Millennium Seed Bank Partnership

mg	Milligram
MIC	Minimum inhibitory concentration
mL	Milliliter
MOA	Memorandum of Agreement
MS	Mass spectrometry
Mtb	Virulent *Mycobacterium tuberculosis* H37Rv
NAPRALERT®	NAtural PRoducts ALERT Database
NBSAP	National Biodiversity Strategy and Action Plan
NIAID	National Institute of Allergy and Infectious Diseases
NOFIP	National Office of Forest Inventory and Planning
NPA	National Protected Area
NR Mtb	Non-replicating virulent *Mycobacterium tuberculosis* H37Rv
NTFP	Non-timber forest products
PAFO	Provincial agriculture and forestry office
PI	Principal Investigator
PL	Pathet Lao
PLM	Palm leaf manuscripts
RFS	Rapid field survey
RLA	Royal Lao Army
RLG	Royal Lao Government
rotavapor	Rotary evaporator
RRC	UIC Research Resources Center
SEAR	Southeast Asia Region
SEM	Scanning Electron Microscope
SI	Selectivity Index
sp.	Species
SPE	Solid phase extraction
TB	Tuberculosis/tubercle bacillus
TM	Traditional Medicine
TMRC	Traditional Medicine Research Center
TMS	Traditional Medicine Station
UIC	University of Illinois at Chicago
USDOS	United States Department of State
WHO	World Health Organization
WWF	World Wildlife Fund

Chapter 1
Introduction and Background

Traditional herbal remedies are routinely used in Laos (also known as the Lao People's Democratic Republic, or Lao PDR). This is due, in part, to the fact that the majority of the population lives in rural settings with limited access to biomedical healthcare. Even when Western medicines are readily available, treatments are commonly supplemented with valued local herbs. Old documents in the form of mulberry paper books and palm leaf manuscripts provide an invaluable record of these treatments through history. Demonstrating their continued presence in Laos for centuries helps to affirm that this art and knowledge of disease treatment is the Intellectual Property of the people of Laos.

With its relatively low population density, Laos holds immense areas of undeveloped forests, containing a wealth of information, medical and otherwise. It is thought by some scientists to be one of the "most botanically unexplored countries in Asia" (Thompson and Thompson 2008; WWF 2012a). The Greater Mekong Sub-region (GMS) includes Cambodia, China, Laos, Myanmar, Thailand, and Vietnam (ADB 2012). There were 519 new plant species in the GMS described between 1997 and 2007, and 145 in 2010 alone (Thompson and Thompson 2008; WWF 2012a, b). However, deforestation is destroying Laos' unique plant diversity at an alarming rate (Stibig et al. 2007). Demonstrating its practical and medical value to what is already there may bring the needed encouragement for conservation and sustainable utilization to take place as outside demands call to clear forests. There are links between environmental harm and poverty (MOIC 2012; UNDP 2012). The 2012 Diagnostic Trade Integration Study produced by the Department of Planning and Cooperation under the Ministry of Industry and Commerce actually noted that "… natural resource development has in some instances actively worsened poverty through resettlement and reduced food security, …" (MOIC 2012). The diversity of plants and animals in Laos are important for cultural identity, which is linked to language, geographical features, and surrounding ecosystems, as well as access to material goods (NBSAP 2004). Many of Laos' diverse cultural groups have an intimate knowledge about the plants in their region.

© The Author(s) 2014
B.G. Elkington et al., *Ethnobotany of Tuberculosis in Laos*, SpringerBriefs
in Plant Science, DOI 10.1007/978-3-319-10656-4_1

Biomedical discoveries may be able to provide another perspective on why natural ecosystems should be left intact.

In its broader context, this research began as part of the Vietnam-Laos International Cooperative Biodiversity Group (ICBG) project. The ICBG aims to improve human health through the discovery of new medicines, biodiversity conservation, and the promotion of scientific research and sustainable economic activity in Vietnam and Laos, focusing on treatments for malaria, tuberculosis (TB), HIV, and cancer (Soejarto et al. 1999, 2009).

Although plants to treat other diseases have been included, the main focus of this research was TB, a disease that is currently ravaging the Asian continent. In 2010, 8.8 million new cases of TB were diagnosed and attributed to 1.4 million deaths (WHO 2012a). Current predictive models estimate that one third of the world's population is infected with latent TB, waiting for the victims' immune systems to be compromised. While TB is a curable disease, it is also a disease that primarily affects people who can't afford the treatments. More than 95 % of TB deaths happen in lower income countries (WHO 2012b). As such, among people with HIV/AIDS, especially in developing countries, TB is a leading cause of death (WHO 2011, 2012b). In Laos, more than 3,800 new cases of TB were identified in 2010, with approximately 5 % of the new cases due to multiple drug resistant (MDR) TB, and 38 % of the patients co-infected with HIV (WHO 2010a).

From drug discovery to patient care, globalization and resistance present complex problems. For example, taking an incomplete regimen of TB medication often leads to MDR-TB. After relocating, foreign-born patients can carry MDR with them to infect people on other continents, further complicating the process of finding treatments that will work (WHO 2012b; CDC 2007; NIAID 2007). Once on a new continent, patients in a minority demographic have to cope both with being a minority, as well as with their illnesses, often increasing stress levels (Campinha-Bacote 2007). Medical ethnobotany and ethnopharmacology offer another perspective that can lend valuable insights into each area.

The primary goal of this research was to find some common ground between biomedicine and traditional healing. Because of the destruction of forests, the degradation of old palm leaf manuscripts, and decreasing interest in traditional medicine by younger generations, it is becoming more and more important to record medicinal plant knowledge before it is lost. This research provides written documentation of some of the medicinal plant knowledge held by the people of Laos. Translating and validating some of the power of traditional medicine used in Laos into biomedical terms through laboratory analyses may serve to demonstrate its importance in a global language. In this case, the translational endeavor was performed through in vitro laboratory analyses with select plant species with a history to treat symptoms of TB. The processes of plant collection, extraction, biological assays, and isolation/elucidation are described in the Biochemical Validation section.

Laos

Because it is associated with an ethnic group, a language, a national identity, and the name of the country, the term "Lao" is rather nebulous (Evans 1995; Grabowsky 1995). It is thus stated here that the term "Laos" will refer to the geographic area delineated by maps, "Laotian" will refer to any person originating from this geographic area, "ethnic Lao" will refer to the ethnic group, and "Lao language" to the official and most widely used language in the country.

Geography and Ecology

Throughout its history, the region of Laos has been an important buffer and trading zone between surrounding powerful states. The Laos of today was officially delineated and mapped by the French in the late 1800s. It is flanked by China to the north, Myanmar (Burma) to the northwest, Vietnam to the east, Cambodia to the south and Thailand to the west, with a land area covering 236,800 km^2, an area slightly larger than the state of Utah (CIA 2013) and about half the size of Thailand. It currently holds 16 provinces: Attapeu, Bokeo, Bolikhamxay, Champasak, Huapan, Khammuan, Luang Nam Tha, Luang Prabang, Oudomxay, Phongsali, Salavan, Savanakhet, Sayabouli, Sekong, Vientiane (Wieng Chan), and Xiang Kuang (Fig. 1.1). Due to a multitude of transliteration systems, many of the province names, along with countless other words in the Lao language, have multiple accepted spellings (Lewis 2009).

Laos lies in a tropical monsoon region with a rainy season from May to November and a dry season from December to April. It is frequently subject to floods and droughts. The terrain is mostly made up of rugged mountains. The tallest mountain, Pou Bia, reaches to more than 2,800 m. In 2005, it was estimated that only around 4 % of the land is used for growing crops (CIA (2013). Though Laos is surrounded by five other countries, it remains isolated by some major geographic obstacles. The Annamite Mountains run along the eastern border between Laos and Vietnam. In the northwest, the Luang Prabang Mountain Range creates obstacles for travel to and from China, Thailand, and Myanmar. The Mekong River runs along the western border, separating much of Laos from Thailand and Myanmar.

The Mekong River is vital to Laos and has been important for travel, fishing, and farming for centuries (Campbell 2009). The word "Mekong" actually comes from the name "Meh Nam Khong," meaning "Khong River" (Campbell 2009). The river originates on the Tibetan Plateau and runs through China, Burma, Laos, Thailand, Cambodia, and Vietnam, before emptying into the South China Sea. It is the earth's tenth longest river, with an estimated length stretching almost 5,000 km, releasing around 475 km^3 of water each year (MRC 2012). With approximately 2.6 million tons of fish produced per year (70 % of which are long distance migrants), it is the largest inland fishery in the world (WWF 2012a). More recently, the river has been

Fig. 1.1 *Map of Provinces in Laos.* Map adapted from Wikipedia (2012). The capital city of Vientiane (Wieng Chan) is marked with the star. The spelling of the provinces varies according to different sources

heavily dammed to produce energy for the surrounding countries, to the detriment of many dependent ecosystems (Baird 2011; WWF 2012a).

Laos belongs to the Indo-Burma Biodiversity Hotspot, signifying that the biodiversity of Laos is extremely rich. The National Office of Forest Inventory and Planning (NOFIP) has listed eight different forest types in Laos, including dry dipterocarp, upper and lower dry evergreen, upper and lower mixed deciduous, gallery, coniferous, and mixed coniferous/broadleaf (Inthakoun and Delang 2011). On a larger scale, these include eco-regions warranting protection from development and include the Northern Highlands, Evergreen Forests of the Annamite Mountains and foothills, Central Indochina Limestone Karst, the Bolaven Plateau, Dry Dipterocarp Forests of the Mekong Plains, and the Mekong River, together with smaller rivers and streams (Nam Theun, Nam Kading, Xe Kong and Xe Banhiang) (Inthakoun and Delang 2011). Along with this, Laos holds four of the World Wildlife Fund Global 200 Eco-regions, which are representative examples of the Earth's diverse natural habitats, critical for biodiversity conservation and global sustainability (Olson and Dinerstein 2002). These include Laos' Annamite Range Moist Forests, Northern Indochina Subtropical Moist Forests, Indochina Dry Forests, and the Mekong River (Olson and Dinerstein 2002). In addition, two wetland areas in southern Laos were deemed as Ramsar sites in 2010. Due to the many ecological roles of wetlands, the Ramsar Convention was established in Ramsar, Iran in 1971, for the protection of wetlands. The treaty commits member countries to the conservation and sustainable use of their wetlands.

Laos is home to some of the richest intact ecosystems on the Indo-China Peninsula. In 1993, the Lao government designated a National Protected Area (NPA) network in an attempt to improve forest management practices and to conserve key habitats and ecosystems. Laos also signed on to the international Convention on Biological Diversity (CBD) in 1996. Article 6 of the CBD calls for the design of a National Biodiversity Strategy and Action Plan (NBSAP) from its signatories. For Laos, the NBSAPs cover environmental protection and conservation, noting that they can reduce poverty and enhance the quality of life and health of its citizens (NBSAP 2004, 2010).

The International Institute for Sustainable Development's 2007 report listed medicinal plants and spices of Laos as "Prioritized for export in the National Export Strategy" (Shaw et al. 2007). Sustainable harvesting may lead to positive economic growth in rural communities. Many predict a potential for "green" niche exports for sectors linking the environment and livelihoods, such as ecotourism, organic farming, silk handicrafts, and medicinal plants and spices (Lazarus et al. 2006; Shaw et al. 2007; Sydara 2007). On the other hand, without adequate planning and regulation, this could have detrimental effects on the environment if not regulated. A major challenge for Laos as development continues will be to enforce environmental regulations, as natural resources must be used sustainably for long-term growth and development.

The government of Laos recognizes the opportunities that non-timber forest products (NTFPs) hold for economic growth, as detailed in the 2012 Diagnostic Trade Integration Study produced by the Department of Planning and Cooperation under the Ministry of Industry and Commerce. Number eight of the National Export Strategy lists medicinal plants and spices. It also notes in the Sectoral Environmental Concerns that there is a need to "Develop and enforce laws and regulations related to the forestry sector as a whole, and especially to NTFPs, such as medicinal plants and spices." Further recommendations were also made to "more scientific and community-oriented management of forest resources; improving statistics on harvesting and exporting; increased awareness of the potential impacts on rural communities...; ensuring that intellectual property considerations are adequately reflected in laws and regulations and improving the quality and quantity of NTFPs; and providing incentives for the private sector to sustainably manage these resources, such as through assurances of benefit sharing of any commercialized products," (MOIC 2012).

History and Culture

Historical events that shape the people of Laos and their medical beliefs and systems date back many millennia. It is accepted that Southeast Asia was settled during multiple separate migrations (Reich et al. 2011). Records of the earliest human habitation in the region of Laos date back at least 5,000 years, with evidence of some of the earliest bronze-making around 3,000 BC (Gorman and Charoenwongsa 1976; Higham 2002). In the example of "Fish Soup" described by Fadiman (1998), a student from Laos gave a lengthy presentation about how to prepare fish soup. The presentation involved many aspects about fishing, including which type of hook to use to catch the fish, where to fish, and finally the herbs and broth used. As illustrated by the student, in Laos no event occurs in isolation. This includes traditional medicine. Separating healing practices from the other historical events would be very difficult and would give a fractured picture.

Khun Bulom and the Tai People

There is a popular story about the origin of the people of Laos and Thailand, with the following version adapted from Tossa (2002), Viravong (1964), and Wyatt (1986). Around 700 CE (CE denotes the Common Era. It is also referred to as AD, beginning in 1 BC in the Julian and Gregorian calendars.), a legend emerged about a certain "Khun Bulom," who gave rise to the Tai people who would later settle all across Asia. According to the myth, a great flood destroyed a foolish group of people who ignored requests from their deity. Only three aristocrats were spared, around what is now known as Dien Bien Phu in Vietnam. The aristocrats came across a gourd, hearing voices inside. One of the aristocrats pierced one end of the gourd with a hot stick, and another cut a hole in the other end with a chisel. People then emerged, with darker skin if coming from the hole that had been burned and with lighter skin if coming from the hole that had been chiseled. The darker-skinned people were known generally as *kha* and were to serve the people who originated from the chiseled hole. The lighter skinned people were to be responsible leaders. As the people prospered and multiplied, the deity saw that they were in need of an overseer, and responded by sending Khun Bulom. Khun Bulom married two wives, with which he had seven sons. These sons became the rulers of the seven major states in Southeast Asia (what are now known as Luang Prabang, Xiang Kuang,

Ayutthaya, Chiang Mai in Thailand, Sipsong Pan Na in China, Hamsavati in Myanmar, and an unclear area in north-central Vietnam).

Some historical accounts hold that the Tai people migrated south and west from Northern Vietnam sometime in the first millennium CE, primarily led by warrior leaders who conquered and assimilated with the indigenous peoples that they encountered (Evans 2002; Stuart-Fox 1997). A conquered village would remain close to the warriors' villages, linking many villages together into a *meuang* or district. Organization by *meuang* is still used in Laos today.

The Tai people adopted many Hindu ideas and beliefs, including that of circular mandalas (Stuart-Fox 1997; Evans 2002). Contrary to the European idea of kingdoms, mandalas stretched from a political center outward. Rulers were less concerned with specific boundaries and were more concerned with networks of people, with mandalas growing and shrinking at different times and with different rulers. Overlapping mandalas were possible and frequent, so that some people were required to pay tribute to multiple rulers (Stuart-Fox 1997; Evans 2002). The mandala state of Lan Na, meaning "a million rice fields," possibly arose around 1200 AD in Northern Thailand, followed shortly after by Sukhothai to its south (Fig. 1.2).

In many historical accounts, the appearance of Fa Ngum in the 1300's marks the beginning of Laos' history as a state (Evans 2002; Lorrillard 2006; Stuart-Fox 1997). Fa Ngum was the leader credited with the capture of *meuang* from Siamese forces to create Lan Xang, which lasted into the late seventeenth century. Buddhist records have been found dating back to the era of Lan Xang (Lorrillard 2006), and a mix of Buddhism and Animism has shaped many of the practices of the people of the region today. When Lan Na fell to Burmese forces, many of the elite fled to Luang Prabang, bringing an influx of Lan Na culture and the famous emerald Buddha with them (Evans 2002).

In the eighteenth century, disagreements over the heir to the throne resulted in Lan Xang dissolving into separate, smaller and weaker kingdoms in Luang Prabang, the Plain of Jars, Vientiane, and Champasak. By 1782, Siam's Ayutthaya had incorporated Vientiane and Champasak, bringing thousands of families as royal servants to Saraburi, an area northeast of Bangkok, and making off with the emerald Buddha (Evans 2002; Stuart-Fox 1997).

Chao Anou (Anuvong), of Lan Xang royal descent, was educated in Bangkok and installed as the vassal king of Vientiane. He tried to reinstate the kingdom of Lan Xang in 1827 by marching on Ayutthaya and made it as far as Saraburi before being pushed back. Chao Anou failed to find enough additional troops to retain control of the region and fled. After the king of Siam learned that Chao Anou had escaped, he demanded that the Siamese army "return to Vientiane and reduce the city to ashes so that Chao Anuvong cannot make use of it" (quoted in Viravong 1964). After destroying Vientiane, tens of thousands of families were rounded up and moved across the Mekong River into Siam's undisputed territory on the Khorat Plateau (Viravong 1964).

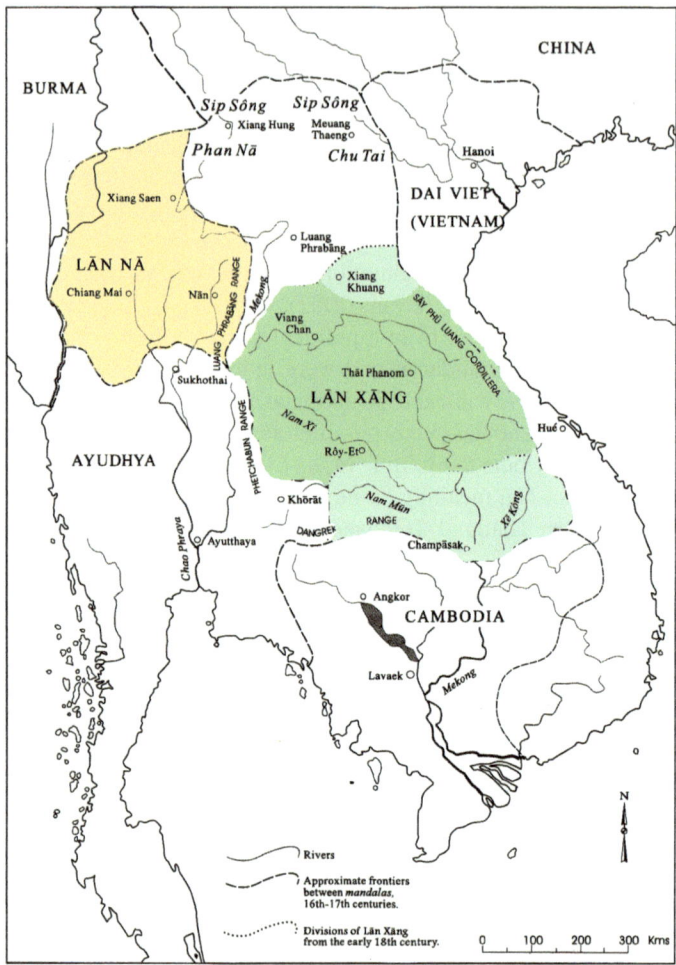

Fig. 1.2 *A depiction of Southeast Asia, sixteenth to nineteenth centuries.* Map adapted from Stuart-Fox (1997) to feature the principalities of Lan Na and Lan Xang. The boundaries of each principality shifted, based on the mandala idea of governance over people instead of geographic boundaries

French Indochina

In the late 1800's, groups of Haw marauders from China began pillaging *meuang* all the way from Luang Prabang to Vientiane (Evans 2002). An account of some of the pillaging is recorded by a British surveyor working for Siam in 1884, who wrote that "the *wats* had been wantonly destroyed, and piles of palm-leaf records lay heaped together, which, unless soon looked at, would be lost forever" (McCarthy

1900). Palm-leaf records of Laos are invaluable religious and historical records. The Siamese could not stop the attacks, and in response, Laos sought help from French-colonized Vietnam (Evans 2002). After an ultimatum from France, Siam ceded all territory east of the Mekong River to colonial French Vietnam in July 1893. This action effectively split the area that had been Lan Xang into halves, with most of the original inhabitants of Lan Xang in Siam's territory. Today there are still more ethnic Lao people in Thailand (known as Isan Thai) than in Laos itself.

World War II and Independence

When World War II broke out and Japan gained control of Indochina in 1945, the Royal Lao Government's (RLG) King Sisavangvong was coerced into declaring Laos' independence from France. However, when Japan surrendered to the Allies later that year, the king revoked his declaration. This led to conflict with other members of the royal family who wanted to retain independence. By 1955, an underground communist organization called the Pathet Lao (PL) had emerged as Laos' equivalent of the Viet Minh and Viet Cong (Vongsavanh 1984; Zasloff and Langer 1970). Though the king himself was not a supporter of the PL, the organization had links in the royal family.

The Pathet Lao and the Secret War

Under the shadow of the Cold War, the next 20 years involved continual political change in Laos, mainly divided into three parts. One part supported the North Vietnamese and Pathet Lao communist movement. Another part supported the constitutional monarchy aided by France and the United States. The third part consisted of those who wished for Laos to remain independent from heavy foreign influence (Stuart-Fox 1997). All three groups struggled financially because the majority of Laos lived as subsistence farmers, and infrastructure problems meant that taxation was extremely difficult. The conflict in Laos came to be almost entirely funded by foreign powers, mainly from the United States and the Democratic Republic of North Vietnam (Evans 2002; Stuart-Fox 1997).

Despite the declaration of Laos neutrality in the 1962 Geneva Accords, North Vietnam continued to use trails through Laos to infiltrate South Vietnam (via the Ho Chi Minh Trail), and the United States covertly fought against North Vietnam by supplying aid to the Royal Lao Army (RLA) and Hmong forces led by Colonel Vang Pao, leading to the "secret war" (Evans 2002; Stuart-Fox 1997). The Hmong were an ethnic minority living in the mountains of Laos. While some ethnic Lao referred to them as barbarians (*meo*), or *kha*, the Hmong did not share a history that could be traced back to Khun Bulom, and did not see themselves as inferior (Evans 2002). Contrary to the more established *meuang* system of the Tai, the Hmong organize themselves by clans and lineages, which has aided their often-nomadic existence. Due to an earlier conflict between two powerful Hmong families, most Hmong groups supported the RLG, and other groups favored the PL (Evans 2002). In 1963, the neutralist troops were conquered, and in 1964, the RLG granted more freedom for the US to fight almost independently, instead of fighting through the RLG, as it had been doing previously (Evans 2002). By 1969, air strikes over Laos had increased to a rate of 200–300 each day (Evans 2002). Remnants of unexploded ordnance continue to pose a threat to the civilian population today. After S. Vietnam fell and the US withdrew in August 1975, the PL declared the liberation of Vientiane. Shortly after, on December 2, King Sisavangvong abdicated the throne (Evans 2002).

The Lao People's Democratic Republic

Laos officially became a one-party state when the Lao People's Democratic Republic (PDR) was declared as the country's new name in December, 1975, led by the Lao People's Revolutionary Party (LPRP). Due to the political instability, by 1980 at least 10 % of the country's population had left the country (Evans 2002). Farming collectivization distanced the peasant population from market opportunities and made them more reliant on subsistence farming. As the country stabilized, different regulations were tested as Laos progressively reopened to the outside world. In 1988, national elections were held by the LPRP. In 1991, a constitution was released. Difficulties with taxation and farming collectivization continued, and by 1994, most state enterprises were privatized.

Despite the outflow of people, Laos remained very ethnically diverse. Based on a French classification from the colonial era, the RLG referred to the population of Laos in three major topographical groups, according to the relative altitude of where they lived: the Lao Loum (lowland Lao, with the majority of the people being ethnic Lao), the Lao Teung (midland Lao), and the Lao Soung (highland Lao, including the Hmong) (Evans 1995, 2002). This simplified classification is still used by many, even though the government of Laos presently recognizes 49 different ethnic groups. In the 2005 census, it was estimated that 55 % of the population is made up of the lowland Lao ethnic group, 11 % Khmou, 8 % Hmong, and 26 % of more than 100 minor ethnic groups. Religiously, 67 % of the population claimed to be Buddhist, 1.5 % Christian, and 31.5 % unspecified. In July 2011, Laos was estimated to have a population of 6,477,211 people (CIA 2013).

Although the abundance of natural resources in Laos has been recognized, it remains one of the poorest countries in Southeast Asia, ranking 138 out of 187 countries on the United Nations Human Development Index (UNDP 2011). However, Laos recently joined the World Trade Organization, and the government has declared its goal to rise above "least-developed country" status by 2020, clearly expecting trade to be the mechanism (Shaw et al. 2007). The trade of medicinal plants and spices is included (MOIC 2012).

Traditional Medicine

> Civilization is on the march in many, if not most, primitive regions. It has long been on the advance, but its pace is now accelerated as the result of world wars, extended commercial interests, increased missionary activity, and widened tourism. The rapid divorcement of primitive peoples from dependence upon their immediate environment for the necessities and amenities of life has been set in motion, and nothing will check it now. One of the first aspects of primitive culture to fall before the onslaught of civilization is knowledge and use of plants for medicines. The rapidity of this disintegration is frightening. Our challenge is to salvage some of the native medico-botanical lore before it becomes forever entombed with the cultures that gave it birth. (Schultes 1963)

The World Health Organization estimates that over a third of the population in developing countries lack access to essential medicines, and that safe and effective traditional remedies may ease this disparity (WHO 2008b). Many Asian countries are underdeveloped, including much of Laos. There are additional reasons to encourage the sustainable use of medicinal plants. Over-harvesting and monoculture can lead to genetic loss, which will impair the opportunities of future generations. The use of medicinal plants can also help reduce the use of wild animal parts, which are lost more rapidly compared to plants (Phiapalath 2009).

Medicinal plants have been used for centuries to treat disease in Laos, and traditional knowledge about the use of these plants has been passed down and is held by many healers today. Many of Laos' diverse ethnic groups have their own complex culture and traditional heritage, including traditional medicine knowledge (Libman et al. 2009). In a 2005 study of six hundred households from lowland and mountainous districts, seventy-seven percent of the families interviewed reported that they used traditional medicine (Sydara et al. 2005).

Along with holding on to this knowledge for cultural reasons, much of Laos has kept medicinal plant knowledge through necessity. Undeveloped forest covers more than forty percent of Laos (ADB 2007; CIA 2013; WorldBank 2006). According to the World Factbook (CIA 2013), around seventy percent of the population lived in rural areas as of 2010, mainly in Mekong River valleys and tributaries. As of 2012, there were an estimated 2.7 physicians for every 10,000 people (WHO 2012a). To put this value into perspective, at the same time, there were nearly ten times the number of physicians per capita in the United States. This puts most of the population into close contact with medicinal plants from forested areas and with limited access to medical clinics and Western medicines. After approximately 820,000 Southeast Asians moved to North America in the 1970s (McCloskey and Southwick 1996), there was a flurry of refugee health-related publications in the US and Canada (Patil et al. 2012). While these published materials deliver health benefits to immigrants, similar studies addressed to the populations that remained in Southeast Asia are lacking (including those in Laos). Consequently, in the event of health problems, it is logically less expensive and easier for these people to turn to traditional herbal remedies than to visit a government clinic.

There has been some debate over whether or not traditional medicinal plants are any more likely than randomly chosen plants to contain compounds that act against microbial targets (Balick 1990; Case 2006; Cragg et al. 1994; Gyllenhaal et al. 2012; Lewis and Elvin-Lewis 1995; Saslis-Lagoudakis et al. 2012). In a review by Gyllenhaal et al. (2012), it was shown that plants that were used traditionally to treat symptoms of tuberculosis in Laos were significantly more likely to yield active test results against virulent *Mycobacterium tuberculosis* H37R*v* (Mtb) than plants chosen at random.

History of Traditional Medicine in Laos

While the origins of traditional medicine in the region are unclear, it is evident that Buddhist and Sino-Indian influences helped shape early traditional pharmacopoeia and practices (Pottier 2007). In the centuries since then, medical traditions have been transmitted both orally through the centuries among different ethnic groups, as well as by way of written documents. These written documents include mulberry paper books and palm leaf manuscripts, usually housed at Buddhist temples. An example comes from neighboring Thailand, which shares many cultural similarities with Laos. University of Pennsylvania professor Maxwell Sommerville visited the area in the nineteenth century, noting that "copies of the principal Pharmacopoeia of Siam are in the possession of the bonzes [monks] in the temples" (Sommerville 2010). To this day, young monks are trained in traditional medicine by elder monks at many of the Buddhist temple-compounds, or *wat*. Information written generations ago about the uses of medicinal plants also appears in palm leaf manuscripts located in the libraries of the larger *wat* (Abhakorn 1997; NLL 2009).

In many traditional medicine systems, plants are believed to reveal their salubrious properties to healers, as in the "Doctrine of Signatures" dating back to the time of Dioscorides. The same belief was held in the late nineteenth-century Thailand, when a medical doctor noted that people "believe that medicine has the power to counteract the deranged elements, and restore them to a healthy equilibrium. The origin and practice of medicine they believe to have been supernatural. Their medicinal books declare that the father of medicine was so privileged that wherever he went, every individual member of the vegeto-medical kingdom was sure to summon his attention, and speak out, revealing its name and medical properties…" (McDonald 1999).

Institute of Traditional Medicine

Due in part to the large percentage of the population relying on traditional medicine, one of the Lao government's primary tasks is to guard public healthcare services through ensuring proper production and use of traditional medicines. The Lao government affirms this through its Institute of Traditional Medicine (ITM). The ITM is primarily concerned with gathering and cataloging traditional treatments, the greater part of which consist of herbal remedies, and with publishing the results of their research for use by contemporary healers and the wider public (Riley 2000;

Soejarto et al. 1999, 2009). The research center was established in 1976 under the Ministry of Health, then known as the Research Institute for Medicinal Plants (RIMP) (1976–1995), and later as the Traditional Medicine Research Center (TMRC) (1997–2010).

As part of its function and mandate, the ITM has established smaller traditional medicine units called Provincial Traditional Medicine Stations (TMSs) (Soejarto et al. 1999, 2009). Initially, a TMS was established in each of ten provinces. Under the support of the ICBG project (1998–2008), three more were added (Soejarto et al. 2009). In 2012, one TMS (Paksan, Bolikhamsai province) was phased out. Today, a total of 12 TMSs is in operation. Each TMS is affiliated with a traditional medicine hospital. The staff of a TMS works in direct contact with rural traditional healers. As a result, the traditional healers are then able to both give and receive information on treatment methods currently in use around the country. The University of Illinois at Chicago (UIC) has been collaborating with the ITM since 1996 in an attempt to study the medicinal plants of Laos and to promote conservation of forested areas holding medicinal plants (Soejarto et al. 2009, 2012). Experts at the ITM have been trained in plant collection and identification, and its phytochemistry laboratory is equipped to extract plant samples. The ITM also holds a collection of medical palm leaf manuscripts.

Linguistics

The rich ethnic diversity of Laos is reflected in its many languages. Based on linguistic history alone, ethnographer Laurent Chazee has divided the people of Laos into 119 different groups (Chazee 1999). The language families can be broken down into the Tai-Kadai (Daic), Austroasiatic (Mon-Khmer), Hmong-Mien (Meo/Miao-Yao), and Tibeto-Burman (Enfield 2006; Lewis 2009; Matisoff 1991). Ethnologue reported 84 individual languages in 2009 (Lewis), giving Laos a very high rate of linguistic diversity in a relatively small area, with the greatest variability in the mountainous areas (Enfield 2006). It is also common to hear Vietnamese, Chinese, French, and English throughout the country. Due to the many different languages and dialects, there has been difficulty in standardizing an official language. Because this research was predominantly involved with plants used by the lowland Lao people who use the Lao language, the following review is also predominantly of the Lao language.

Spoken

The spoken Lao language falls under the Tai language umbrella, which relates to the group of people originating from southern China (Enfield 2003; Hartmann 2002). Today, people speaking Tai languages form the majority of Laos and Thailand. Thai and Lao languages are essentially dialects (Enfield 2006), and as such, are very similar. The spoken Lao language is primarily used by the people inhabiting the lowlands near rivers, which include more than two million people in Laos and another twenty million people in northeast Thailand (in Thailand, the language is typically referred to as "Isan") (Enfield 2006; Lewis 2009). The Lao language itself has differing dialects and vocabularies in different regions. It is a monosyllabic tonal language, and the written characters of the Lao script directly indicate the tone. Many scholars have different systems for classifying the different tones (Hartmann 2002).

For religious ceremonies, Pali is an Indic language, believed to be closely related to the language used for teaching by Gautama Buddha (Hazra 1994). Although it is technically classified as an extinct language (Lewis 2009), it is still chanted in many Theravada Buddhist ceremonies. It is the language of the earliest Buddhist scriptures and has been transcribed with a variety of written scripts, including roman, the International Phonetic Alphabet, and many Southeast Asian scripts. In Laos, Pali can be found frequently in religious documents, including the palm leaf manuscripts.

Written

As seen previously, more than half of the population of Laos is made up of different ethnic minorities, each with their own methods of recording history. Because the ethnic Lao are primarily Buddhist and have kept written records in the country dating back to the sixteenth century (Hundius and Wharton 2011), the vast majority of the old medical documents surveyed in this research represent this group of ethnic Lao Buddhists.

The Lao script contains 33 consonants and 28 symbols for vowels. It is written and read from left to right, but the vowel symbols may appear before, around, and/ or after the consonants. Traditionally, there are no spaces between words. The Lao language is phonetic, and each combination of consonants, vowels, and tone marks represent a specific tone and sound that remain constant.

Translating the Lao language into English can be complicated because of the lack of an official transcription system. As a result, there are innumerable different English spellings of Lao words. For example, one of Laos' western provinces has been spelled "Sainyabuli" (Cummings and Burke 2005; NLL 2009), "Saiyabouri" (Pholsena and Banomyong 2006), "Xaignabouli" (CIA 2013), "Xaignabouri" (Evans 2002), and "Xayabuly" (ADB 2009), among other variations in major works about Laos. Another example of a transcription complication is in the letter "R," which was used conventionally in some Lao words, then dropped for a few years, and is once again used. For example, *Luang Prabang* was officially changed to *Louang Phabang* in 1995 (Enfield and Evans 2000; Hartmann 2002), but now appears as *Luang Prabang* in most documents.

As opposed to the every-day written Lao script, the Tham script is used for most religious texts in Laos (McDaniel 2008). The word "Tham" signifies the term for Dhamma in the Pali language. It is a written script derived from a Mon alphabet, which, in turn, has roots in South India. Usually only highly trained monks in the region are instructed in reading and writing Tham. The rounded characters found in the palm leaf manuscripts are probably indicative of the materials used in their creation; straight lines and angles tend to tear palm leaves, while curved lines do not. Conversely, rectangular characters may be associated with records carved in stone (Hartmann 1986).

Medical Manuscripts

Mulberry Paper Books

One type of material that has been used to keep records is made from the bark of the paper mulberry tree (*Broussonetia papyrifera* (L.) L'Hérit. ex Vent., Moraceae), known as *sa* in the Lao language (Fig. 1.3). Paper is made from the inner bark. The process consists of pounding or boiling the bark until it is soft, followed by soaking it with alkali until the fibers are pliable. The fibers are then pounded until they separate, then mixed with water and homogenized. The mixture is then spread over a screen and allowed to settle, and the resulting pulp is dried. Using mulberry paper allows the creation of a large, smooth surface area to write on.

B. papyrifera is native to East Asia. It is a deciduous tree with leaves that vary in shape, ranging from ovate to deeply lobed. Similar to other mulberry trees, the fruits are very sweet. Additional taxonomic information is available in the *Flora of China* (Flora of China 1994a).

Palm Leaf Manuscripts

The use of palm leaves for keeping records is believed to have originated in India, dating back to the sixth century BC and was brought to Laos by Buddhist missionaries in the form of Pali-Sanskrit scriptures (Hartmann 1986). Traditionally, Buddhist monks have created palm leaf manuscripts as a means of keeping religious records. However, the manuscripts have also been used to keep other kinds of records, including history, law, customs, astrology, and magic, as well as traditional medicine and healing.

The palm leaf manuscripts are typically kept in Buddhist temples to serve as references for monastic and lay-people and are regarded as sacred. However, with the increasing availability of paper books, many of the palm leaf manuscripts have fallen into disuse and neglect, leaving them subject to insects, mold, and other destructive forces. This has been recognized in recent years, and an inventory project was launched in the 1980s under the leadership of Dara Kanlaya and funded by the Toyota Foundation (Hundius 2005). Subsequently, in 1992, the Lao-German Preservation of Lao Manuscripts Programme (PLMP) was created and built on the

Fig. 1.3 *Mulberry paper leaf and book.* (**a**) The bark of the paper mulberry tree, *Broussonetia papyrifera* (L.) L'Hérit. ex Vent. (Moraceae), has been used to make paper in Laos. (**b**) The paper mulberry book in the picture contains entries about medical treatments and is kept at the National Library of Laos

Toyota Foundation project in order to preserve the manuscripts and to record them on microfilm. Through the Lao Ministry of Information & Culture and supported by the German Ministry of Foreign Affairs, the PLMP surveyed more than 800 *wat* and has preserved approximately 86,000 texts (Hundius 2005).

Through active interaction with local communities, the project has emphasized the importance and proper handling and storage techniques of the manuscripts. During the surveys, microfilms were recorded in situ from about 12,000 texts. Each text in the microfilm was recorded with a unique code, allowing the origin and the subject material of the manuscript to be easily recognized. These microfilms are now incorporated in the Digital Library of Lao Manuscripts Collection (DLLM), the majority of which were made from palm leaves. The National Library of Laos has a website allowing the public to access more than 8,000 of the images on micro-film (Hundius and Wharton 2011; NLL 2009).

Creation

The palm leaf manuscripts of Laos are typically created from palm leaves. These reportedly include the talipot palm (*bai Lan*; *Corypha umbraculifera* L., Arecaceae) or the Asian palmyra palm (*bai Tan*; *Borassus flabellifer* L., Arecaceae). *C. umbrac-ulifera* is native to India, but has spread naturally to other regions of Southeast Asia and has been introduced to countries beyond, while *B. flabellifer* is native to a wider region of India and Southeast Asia (de Zoysa 2000). The leaves of the talipot palm are more durable and easy to work with, but difficult to find in Laos. The fan-shaped leaves of both species are borne at the tip of a thick stalk (petiole), from which pleated, veined leaf segments emanate outwards, each terminated by a sharp blade.

The palm leaves used to make the manuscripts are very thick and durable and can last for hundreds of years. The leaves are pressed, dried, and cut into uniform shapes, usually around 10 cm in width and ranging anywhere from 15 to 60 cm in length. Then a sharp object, like a metal-tipped stylus, is used to scratch characters onto the leaves. Usually, there are four or five lines of engraved writing on each side. Because no color is used with the stylus, oil and soot are then rubbed onto the leaves. The soot gets caught in the scratches to make the characters stand out.

The leaves are threaded together with cord and often held between wooden covers and are usually wrapped in cloth for storage. Although they are susceptible to decay by mold and insects, if the manuscripts are properly stored, they will last for hundreds of years. One can often tell the subject material of manuscripts apart simply from the length of the bundles. For example, a religious manuscript will be the length of an arm, while something used on a daily basis, such as medical manuscripts, are usually the length of a forearm. See Fig. 1.4.

At least seven different spoken languages have been found in the manuscripts held at the National Library of Laos, transcribed in at least nine distinct written scripts (NLL 2009). However, in the majority of the manuscripts in Laos, the Tham script is used to transcribe the spoken Lao language (McDaniel 2008; NLL 2009). This could be likened to the idea that the spoken English language can be written in Vietnamese or Japanese script, but the Latin/Roman script is usually used.

While the manuscripts are usually religious in nature, many Buddhist monks have also been healers and have kept a record of their treatments in the palm leaf and mulberry paper manuscripts (Pottier 2007; Tiyavanich 2003). Some of the man-uscripts containing information about healing and medicine date back at least 200 years. The National Library has digitized at least 1,500 microfilm images contain-ing information about healing or traditional medicine. Adding to the data, the ITM in Vientiane holds a collection of medical manuscripts and began translating them into modern Lao script more than 10 years ago. For the most part, each bundle of medical manuscripts contains treatments for diverse afflictions instead of being grouped by specific diseases (Elkington et al. 2012). One page may include a

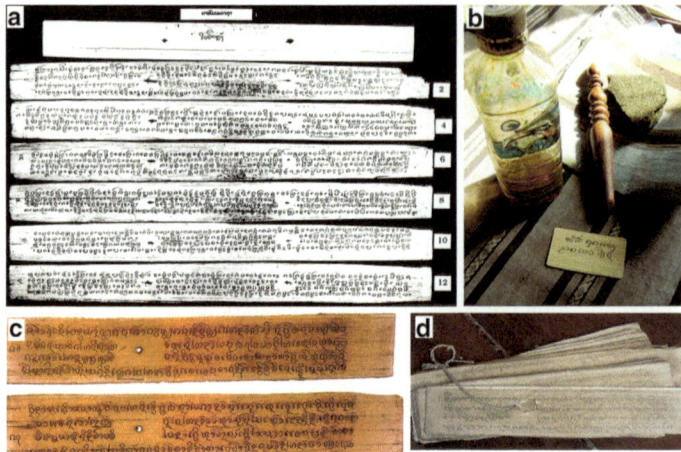

Fig. 1.4 *Palm leaf manuscripts*. (**a**) Digital image of microfilm adapted from the National Library of Laos. www.laomanuscripts.net. (**b**) Tools used in writing on the palm leaves, including a mixture of oil and soot in a plastic bottle, a wooden stylus, and a sponge. (**c**) Medical palm leaf manuscripts. (**d**) Medical palm leaf manuscripts strung together

description for treatments of skin rash, headache, and coughs, while the next page is dedicated only to treatments for cough. There may even be the odd entry for a love potion or how to regain lost items thrown in. However, the vast majority of medical manuscripts analyzed for this research were about the use of different plants to treat different ailments.

Tuberculosis

Robert Koch first discovered and isolated what he named "tubercle bacilli" in 1882, later named *Mycobacterium tuberculosis*. Before it was as treatable as it is today, tuberculosis was referred to as "consumption," as it seemed to consume its victims from the inside (Koehler 2002). It has also been referred to as the white plague because of the white pastiness of a victim's skin (Kent 1992). It has since been found that these mycobacteria evolved with a proposed origin in East Africa almost three million years ago and may have afflicted some of the early humans (Gutierrez et al. 2005). Narrowing down to TB in Asia, it was actually documented in a 500,000-year-old buffalo in China. A human female with marks from TB on her spine was buried in NE Thailand in the Iron Age (Tayles and Buckley 2004), and the disease was prevalent enough to make an appearance in China's popular literature in the 1790s CE, when the heroine of Dream of the Red Chamber, Lin Dai-yu, was plagued by TB (Xueqin 1974, reprint).

Mycobacteria lack an outer cell membrane and are grouped with gram-positive Actinobacteria (Ventura et al. 2007). The Latin prefix "myco—" means both fungus and wax, relating to the "waxy" compounds in the cell wall. Virulent *Mycobacterium tuberculosis* H37Rv (Mtb) specifically has a relatively slow division time of 15–20 h. To put this into perspective, some *E. coli* strains divide in only 20 min. This is important because most TB drugs act during the division process, meaning that they must be taken for a longer time if all of the bacteria are to be killed. Presently, the first-line treatments for tuberculosis include isoniazid, rifampin, ethambutol, and pyrazinamide (Merck 2009; NIAID 2010). A typical treatment consists of 2 months for all of these drugs, with an additional 4 months of isoniazid and rifampin (Merck 2009; NIAID 2010). Mtb is an obligate aerobe, is non-motile and its natural reservoir is in the human body.

TB is usually acquired when a person inhales Mtb that have been exhaled or coughed out from an infected person. The inhaled bacteria then go to the lungs, where they are taken up by macrophages. Around 5–10 % of the people exposed to pulmonary TB become infected (WHO 2010b), and those who are infected but do not get sick have non-contagious latent, or non-replicating (NR) TB, where the bacteria are dormant and inactive, but still alive. Around 10 % of the people who are infected develop the disease over their lifetime (NIAID 2010).

The current drug regimen is very effective for drug-sensitive TB if the patient complies with the treatment. However, the treatment is time-consuming and often leaves the patient with uncomfortable side effects. Patients often take the drugs only until the major symptoms clear up, which could be in as little as a few weeks. Mycobacteria that have been exposed to the drugs, but not killed by them, are likely to become resistant to the drugs. This has contributed to the emergence of MDR and extensively drug-resistant (XDR) TB (NIAID 2010). MDR TB occurs when there is resistance to isoniazid and rifampin, requiring a new combination of drugs. These second-line drugs often have increasingly uncomfortable side effects, the treatment may last much longer, and may cost much more than first-line therapy drugs (NIAID 2007). For example, the first TB drug to be published, streptomycin, has nerve toxicity. In addition, as with all drugs used for treating active TB, mycobacteria develop resistance to it relatively quickly, requiring it to be administered with a number of other drugs (Murray 2004).

As a way to help developing countries to tackle TB, in the early 1990s the WHO launched a five-part program called Directly Observed Treatment Short Course (DOTS). The first two parts entail ensuring that countries have sufficient resources to correctly diagnose and treat TB; the third part encourages patients to comply with their treatments; the fourth seeks to ensure that there are enough drugs to treat everyone; and the fifth part is to make sure that the records and reports are accurate (WHO 2006).

International Collaborative Research

There is understandable concern about traditional medical knowledge from less-developed countries being used to generate revenue for foreign pharmaceutical companies, while the communities who provided the information are weakly compensated or not compensated at all. While the CBD has increased awareness about ownership of genetic materials and intellectual property (UN 1992), specific guidelines for how to address these issues remain vague. The more recent Nagoya Protocol has lessened some concerns by providing a more clear framework for the benefit sharing as a result of the utilization of genetic resources (UN 2011).

While other countries in the region have had more opportunities to discuss steps for developing a legal framework flexible enough to protect traditional knowledge and practices and still foster research, Laos has only recently begun to develop laws aiming to protect its vast biodiversity resources. With a goal to fairly acknowledge and compensate contributions derived from traditional medicine knowledge of Laos used in this project, this project was also designed to follow the guidelines set down more recently by the International Society of Ethnobiology (ISE).

The ISE Code of Ethics acknowledges the strong ties between culture, language, land, territory, and biological diversity. It lays down 17 principles covering Prior Rights and Responsibilities, Self Determination, Inalienability, Traditional Guardianship, Active Participation, Full Disclosure, Educated Prior Informed Consent, Confidentiality, Respect, Active Protection, Precaution, Reciprocity, Mutual Benefit and Equitable Sharing, Supporting Indigenous Research, The Dynamic Interactive Cycle, Remedial Action, Acknowledgement, and Due Credit and Diligence (ISE 2006).

Thus, this research project was performed under a Memorandum of Agreement (MOA) established between the ITM and UIC, detailing the objectives, responsibilities, and benefits of the involved parties. Based on the MOA for the Vietnam-Laos-UIC ICBG project, the core components address intellectual property rights, prior informed consent, and a benefit-sharing plan. These components are separated into eleven parts, covering academic exchanges; joint research; UIC, TMRC/ITM, and joint responsibilities; intellectual property rights; biological material transfer; dispute resolution; and renewal and amendment, among other things. The terms and conditions of the entire ICBG project are covered in more detail by Soejarto et al. (2004). A copy of the MOA has been published by Elkington (2013). The research protocol, which involved interviews with human subjects, was approved by the UIC Institutional Review Board (UIC-IRB) (#2007-0396).

Through this research, medicinal plant knowledge held by the people of Laos was recorded and documented. By demonstrating that treatments have been present in Laos for centuries, this study has also helped to affirm that the unique art and knowledge of disease treatment used in Laos is the intellectual property of the people of Laos. Information acquired from this research about traditional herbal remedies and/or plants is being shared with the communities in Laos who may benefit from the knowledge, helping healers to refine their treatments and linking together healers from different regions of the country. As its ultimate goal, such information is intended to safeguard traditional medicine knowledge from being lost by creating written records that can be shared with younger generations.

Chapter 2
Ethnobotany

Manuscript Analysis

It is to be remembered that this research focused on tuberculosis (TB), and in a rural setting, it would be unlikely to concretely diagnose TB, as opposed to other respiratory ailments. The traditional treatments encountered, therefore, were sought according to TB's main symptoms. Through the National Library's categorization methods, 31 rolls of microfilm containing 1,039 digital manuscript images were identified as containing information about healing practices or traditional medicine. Translators manually read through all of these pages of digital manuscripts and made a list of what they found. Simultaneously, other translators went through six actual bundles of medical manuscripts held at the Institute of Traditional Medicine, the National Library of Laos, and at a few temples, also making a list of the diseases that they found. The lists were recorded in an Excel spreadsheet, giving the location of each entry to create an index to the manuscripts. Thus, in the future, one can simply look up a desired key term and then go to the actual manuscripts to learn further details about the treatments. Presently, the database contains 9,706 entries from the manuscripts. Of these, there are 325 disease entries that have come from actual palm leaf and/or mulberry paper manuscripts and 9,381 entries from the microfilm. There were 1,210 citations of *sainyasat*, very roughly translated as "magic" (McDaniel 2011), 711 entries that did not contain medical information, and 532 citations of entries that were unclear or illegible. Of the 9,706 total entries in the database, 7,080 citations give information on treatment of specific ailments (Elkington et al. 2012).

The palm leaf manuscripts that were studied originated from Champasak, Khammuan, Luang Prabang, Salavan, Savanakhet, Sayabouli, and Vientiane provinces. Four written scripts (Lao Buhan, Tham Lao, Tham Leu, and Thai) were used to transcribe four spoken languages (Lao, Pali, Tai Leu, and Thai). Only three of the manuscripts had any indication of a date when they were created: 1275 CS (1913 CE), 2491 BE (1948 CE), and 2520 BE (1977 CE). CS stands for "Cullasakarāja,"

© The Author(s) 2014
B.G. Elkington et al., *Ethnobotany of Tuberculosis in Laos*, SpringerBriefs in Plant Science, DOI 10.1007/978-3-319-10656-4_2

which is the era beginning at 638 CE in Lan Xang and Lan Na. CE stands for the "Common Era" or the "Christian Era." This is the same as "Anno Domini" or AD. BE stands for the "Buddhist Era," beginning in 545 BC or BCE (Before Christ, or Before the Common Era) in Thailand and Laos. All of the other manuscripts were undated. As described previously, most of the manuscripts surveyed in this project were written in the Lao language and transcribed using the Tham script.

Treatments containing the term "fever" (*khai*) were listed quite frequently, with 1,054 citations. This was expected as a fever may be indicative of many different ailments. For example, a search for "fever" on the World Health Organization website brings up links to more than 14,600 entries (WHO 2012c). Similar studies have found that complaints of a fever are often treated traditionally in Laos (Elkington et al. 2010, 2014a). Pain was indicated by at least two different terms (*jep* 433; *pouat*-131) and the term for "head" (*hua*) came up 326 times. It is interesting to note that there were 74 entries containing "hiccups" (*sa eu*), but only seven occurrences of the term for "liver" (*tap*), a major cause for discomfort in Laos today. In fact, liver cancer is the leading cancer among people of Laos (WHO 2008a), with lower survival rates than those seen in any other ethnic groups globally (Kwong et al. 2010).

To find entries specifically linked to TB, the full entries related to respiratory ailments were then translated into the Lao and English languages. Key search terms for those entries included tuberculosis, cough, asthma, and lung, which were found in 315 entries. These entries were translated from the medical manuscripts into English, giving a list of 1,133 plants based on their common names. Taxonomic identity of the common names was inferred from ITM experts and from published works by Vidal (1959) and Inthakoun and Delang (2011). Plants that were collected during follow-up field work were then identified as described in the section below (Healer Interviews). From the list of 1,133 plants, 187 plants were repeated more than once. The most frequently listed plants are given in Table 2.1.

See the Glossary at the end for the Lao script of the plant names. These were the only plants listed seven or more times in the manuscripts for treating symptoms of TB. A comparison of the plants most commonly reported by healers is given in Table 2.2.

Difficulties encountered in this segment of the project were mainly in spelling and translation. For example, the previously mentioned term *sainyasat*, which can be spelled two different ways in the Lao script (McDaniel 2011) and other culture-bound syndromes involving spiritual and Animist terms, such as the *khwan*, *Phii*, and *lom*, are rather ambiguous and difficult to translate into English biomedical terms (Pottier 2007). Additionally, many entries were non-specific. The entry *fii nai thong*, roughly translated as "sores in the abdomen," was translated into "tuberculosis" by one translator and "stomach ulcers" by another.

Table 2.1 Plants most frequently cited for respiratory ailments in the palm leaf manuscripts

Scientific name	Common name	Plant part	Number of citations
(Zingiberaceae) *Zingiber officinale* Roscoe	*Khing*	Rhizome	54
(Solanaceae) *Capsicum annuum* L.	*Phik*	Fruit	36
(Poaceae) *Saccharum officinarum* L.	*Oy Dam*	Stem	23
(Amaryllidaceae) *Allium sativum* L.	*Ka Thiem*	Bulb	22
(Poaceae) *Oryza sativa* L. var. *dura* Crevost & Lem.	*Kao Jao*	Seed	19
(Menispermaceae) *Tinospora crispa* (L.) Hook. f. & Thomson	*Kheua Khao Hor*	Stem	17
(Rutaceae) *Clausena harmandiana* (Pierre) Guillaumin	*Song Fa*	Root	16
(Poaceae) *Imperata cylindrica* (L.) Raeusch.	*Nya Kha*	Aerial parts	15
(Moraceae) *Streblus asper* Lour.	*Som Phor*	Stem	14
(Myrtaceae) *Decaspermum fruticosum* J.R. Forst. & G. Forst.	*Khii Lek*	Stem	11
(Plumbaginaceae) *Plumbago indica* L.	*Pit Phii Deng*	Root	11
(Fabaceae) *Acacia leucophloea* Willd.	*Som Phory*	Bark, fruit, root	11
(Acanthaceae) *Thunbergia grandiflora* Roxb. or *Eranthemum pulchellum* Andrews	*Nam Neh*	Root	10
Not Identified	*Bou Ra*	Leaves	10
(Solanaceae) *Solanum cyanocarphium* Blume or *S. melongena* L.	*Kheua Kheun*	Fruit, root	9
(Zingiberaceae) *Curcuma parviflora* Wall.	*Khing Kheng*	Rhizome	9
(Rubiaceae) *Hymenodictyon orixense* (Roxb.) Mabb.	*Som Kop*	Root	9
Not identified	*Ho La Dan*	Whole plant	8
(Bignoniaceae) *Millingtonia hortensis* L. f.	*Kang Khong*	Root, stem	7
(Asteraceae) *Eclipta prostrata* (L.) L.	*Nya Hom Keo*	Whole plant	7
(Poaceae) *Eleusine indica* (L.) Gaertn.	*Nya Phak Khouay*	Whole plant	7
(Verbenaceae) *Vitex trifolia* L.	*Phii Seua*	Leaf	7
(Apocynaceae) *Aganonerion polymorphum* Pierre in Spire & A. Spire or *Ecdysanthera rosea* Hook. & Arn.	*Som Lom*	Root	7
Not identified	*Kweng*	Leaf, root	7

Healer Interviews

The process of working with the indigenous traditional knowledge of contemporary healers started at the government level, by setting down an MOA between the University of Illinois at Chicago (UIC) with the Institute of Traditional Medicine

Table 2.2 Plants most frequently reported by healers for treating respiratory ailments

Scientific name	Common name	Plant part	Number of reports
(Asteraceae) *Elephantopus scaber* L.	*Khii Fai Nok Khoum*	Whole plant	12
(Irvingaceae) *Irvingia malayana* Oliver ex Bennett	*Bohk*	Stem	11
(Bignoniaceae) *Millingtonia hortensis* L. f.	*Kang Khong*	Stem	11
(Solanaceae) *Solanum cyanocarphium* Blume or *S. melongena* L.	*Kheua Kheun*	Root, stem	11
(Zingiberaceae) *Zingiber officinale* Roscoe	*Khing*	Rhizome	8
(Euphorbiaceae) *Chaetocarpus castanocarpus* (Roxb.) Thwaites	*Bohk Khai*	Stem	5
(Molluginaceae) *Glinus oppositifolius* (L.) Aug. DC.	*Mak Kheng Khom*	Aerial parts	5
(Lauraceae) *Litsea cubeba* (Lour.) Pers.	*Sii Khai Tohn*	Stem	5
(Fabaceae) *Mucuna pruriens* (L.) DC.	*Tam Yay*	Root	5
(Rutaceae) *Melicope pteleifolia* (Champ. ex Benth.) T.G. Hartley	*Khom La Wan Joh*	Stem	4
(Lygodiaceae) *Lygodium microphyllum* (Cav.) R. Br.	*Koot Ngong*	Aerial parts	4
(Bignoniaceae) *Oroxylum indicum* (L.) Kurz	*Lin Mai*	Stem	4
(Poaceae) *Cymbopogon nardus* Rendle	*Sii Khai*	Whole plant	4
(Smilacaceae) *Smilax glabra* Roxb.	*Ya Hua*	Rhizome	4
(Fabaceae) *Crotalaria pallida* Aiton	*Hing Hai*	Aerial parts	3
(Malvaceae) *Mansonia gagei* Drummond or (Rubiaceae) *Tarenna hoaensis* Pitard	*Jwang Hom*	Leaf	3
(Solanaceae) *Solanum lasiocarpum* Dunal	*Mak Euk*	Stem	3
(Poaceae) *Saccharum officinarum* L.	*Oy Dam*	Stem	3
(Araliaceae) *Heteropanax fragrans* Seem.	*Oy Xang*	Root, stem	3
(Rutaceae) *Glycosmis pentaphylla* (Retz.) DC.	*Som Xeun*	Stem	3

(ITM, a daughter institute of the Ministry of Public Health of Laos). The purpose and specific nature of this MOA are described above in the section on International Collaborative Research. In general, the ITM would contact the provincial level Traditional Medicine Stations (TMS) prior to a field trip, and the TMS would in turn contact village chiefs, abbots, and/or healers to ask about their availability and willingness to be interviewed. The field interviews for this research were conducted in the provinces of Bokeo, Bolikhamxay, Champasak, Luang Prabang, and Vientiane.

When the research team actually arrived at the TMS, the TMS Head received the research team to discuss the work plan. Accompanied by the TMS Head, the research team traveled to the village where the healer was located, first stopping to meet with the head of the village to obtain clearance. Together with the Village Head and the TMS Head, the team then met with the healer, usually at their home. After initial introduction by the head of the village, the healer was provided with a

Prior Informed Consent (PIC) sheet in the Lao language, describing the research and what the healer's part would be if he/she consented to the interview. A copy of the PIC sheet is available upon request (Elkington 2013). In light of the relatively high frequency of illiteracy in rural areas, the healer would also be told about the project orally in the Lao language, in which case consent to proceed with the interview was verbal. The interviewer(s) would then ask the healer questions following a semi-structured interview guide. The questions were loosely based on those used by Kleinman et al. (1978), to give an indication of how different diseases are perceived traditionally.

Part of the interview process involved the healer describing the location where certain plants could be found. At that time, the research team often walked into a garden or a forested area where the healer would point out the plant(s) in question. Under the permit granted by the Ministry of Agriculture and Forestry of Laos, herbarium specimens were collected, the plants were photographed, and GPS coordinates and field notes were recorded. Attempts were made to collect flowering and/or fruiting specimens to allow accurate taxonomic identification. The section on Taxonomic Documentation describes the details of plant collection methods, collection permits, voucher herbarium documentation, and taxonomic identification in more detail.

The healer interviews were conducted in five provinces: Bokeo, Bolikhamxay, Champasak, Luang Prabang, and Vientiane. Healer age ranged from 29 to 85 years, averaging out at 56 years. There were 10 female healers and 48 male healers. Ethnically, 7 healers claimed to be Hmong, 37 Lao/Lao Loum, 1 Lao Leu, 4 Moy, 2 Nyaw, 2 Phutai, 2 Phuan, 1 Tai Dam, 1 Tai Daeng, and 1 Tai Maen. When asked about religious/spiritual beliefs, 8 healers claimed to be Animist, 45 Buddhist, 1 both Buddhist and Animist, and 3 Catholic. There were seven healers who actually were Buddhist monks or nuns. Professionally, farming provided the mainstay for 23 of the healers. For the others, 19 were civil servants, with 16 of these working in healthcare professions. There were eight healers who used traditional healing as their main profession, with four being traditional medicine vendors and four being healers exclusively. Formal education levels ranged from zero years to university degrees.

In general, the healers claimed to treat stomach ailments most frequently, followed by yeast infections, diabetes, fevers, joint pain, skin problems, kidney problems, beriberi, cancer, colds, coughs, diarrhea, malaria, problems with nerves, back pain, dengue fever, and urinary tract infections. For respiratory ailments, three healers claimed to have treated only two patients for a cough, and another claimed to have treated at least 1,000 patients for coughs. Each healer had a unique arsenal of plants used to treat many complaints, but *Cyclea barbata* Miers (*Mor Noy*) and *Tiliacora triandra* (Colebr.) Diels (*Yanang*), both belonging to the family Menispermaceae, were the most frequently reported species to have the capability of treating many different affections.

When asked where they learned how to treat patients in their traditional practice, 4 healers said they had learned from books, with 5 more saying they learned specifically from palm leaf manuscripts, 18 learned from family, 11 learned from other healers (non-family), 2 from their own experience, 6 learned from Buddhist temples, and 1 healer claimed to receive messages from spirits that appeared in dreams.

When asked what TB is, 1 healer said that it has to do with the heart and 43 other healers said that it had to do with the lungs (primarily "dry lungs" and "lung disease"). The most popular other names for TB included "asthma," "cough," "dry cough," "dry lungs," and "fluid in lungs." When asked about the main symptom of TB, the term for "cough" came up 27 times, "phlegm" eight times, "blood" six times, "throat" four times, and "weight" one time. When asked how a person gets TB, 22 healers responded that it comes from being near a person who has it (sleeping near, being coughed on, sharing food/utensils) and one that it comes directly from eating dirty food. When asked what they would do first if they themselves or someone in their family were to get TB, 26 healers said they would treat it with traditional medicine, four said they would seek help from a hospital, one said to go to the hospital and to use traditional medicines simultaneously, and three said first to quarantine the patient.

A list of 341 plants resulted from the 58 interviews with contemporary healers specifically for treating symptoms of TB. Of these, 165 were named more than once. The most frequently cited plants are listed in Table 2.2.

See the Glossary for the Lao script of plant names. These were the only plants reported by three or more healers in this research. A comparison of the plants most commonly reported in the PLM is given in Table 2.1.

Correlation of Interviews and Manuscript Entries

It was interesting that spiritual healing was only mentioned in one interview (in which the healer received information in his dreams), but came up frequently in the palm leaf manuscripts. This could be because the team conducting the interviews seemed to have western-influenced backgrounds, so perhaps the healers gave responses that they thought would be better understood by the researchers. It was unfortunate that more time could not be spent with each healer and that the need for some standardization required the semi-structured interview survey. In general, all of the healers seemed more at ease when walking in the forest and pointing out the plants and less at ease with the written survey form. While the importance of surveying both males and females has been demonstrated (Pfeiffer and Butz 2005), this study made no preference as to the gender of the healer, and the majority of the healers chosen by the TMS and heads of each village happened to be male. It also may have been more beneficial to spend extra time looking for healers that specialized in respiratory ailments instead of interviewing all of the healers. For example, the older healer who claimed to have treated more than 1,000 patients for cough may have had additional interesting data to share if more time had been allotted.

To identify the most widely used plants through time, lists of the plants named by traditional healers (341 plants) and those written in the palm leaf manuscripts (1,133 plants) were created. The two lists were combined, resulting in a larger list of 1,474 plants. Plants were then ranked (i) according to how frequently they were named, (ii) if they were reported in more than one region, and (iii) if they were cited both in the manuscripts and by healers. This resulted in a list of ten highly ranked plants.

A point system was utilized to prioritize plants that are frequently used in different places and chronicled by different sources. Three categories were used, namely, (a) citation frequency, (b) citation distribution (province where the plant was named), and (c) citation source (healer and/or manuscripts). Each plant was given one point for each citation. Thus, to account for plants used in many areas, one point was given for each province where a plant was cited. To predict which plants have been used over longer spans of time, one point was given for citation in the manuscripts, and one for citation by healer. The three values were then multiplied to give the weighted score. See Table 2.3. An analysis of the geographic origin of the ten most commonly reported plants in both the PLM and healer interviews is shown in Fig. 2.1.

Table 2.3 Ten most frequently cited plants

Scientific name	Common name	PLM entries	PLM provinces	Healer citations	Healer provinces	Weighted score
(Zingiberaceae) *Zingiber officinale* Roscoe	*Khing*	54	Champasak, Luang Nam Tha, Luang Prabang, Sayabouli, Savanakhet, Vientiane	8	Champasak, Vientiane	744
(Poaceae) *Saccharum officinarum* L.	*Oy Dam*	23	Champasak, Luang Prabang, Sayabouli, Savanakhet, Vientiane	3	Bolikhamxay, Champasak	312
(Solanaceae) *Solanum cyanocarphium* Blume or *S. melongena* L.	*Kheua Kheun*	9	Champasak, Luang Nam Tha, Sayabouli, Savanakhet, Vientiane	11	Bolikhamxay, Champasak, Vientiane	240
(Menispermaceae) *Tinospora crispa* (L.) Hook. f. & Thomson	*Kheua Khao Hor*	17	Luang Prabang, Sayabouli, Savanakhet, Vientiane	2	Champasak, Vientiane	190
(Rutaceae) *Clausena harmandiana* (Pierre) Guillaumin	*Song Fa*	16	Champasak, Sayabouli, Salavan, Savanakhet, Vientiane	3	Champasak	190

(continued)

Table 2.3 (continued)

Scientific name	Common name	PLM entries	PLM provinces	Healer citations	Healer provinces	Weighted score
(Asteraceae) *Elephantopus scaber* L.	*Khii Fai Nok Khoum*	6	Sayabouli, Salavan, Vientiane	12	Bolikhamxay, Champasak, Vientiane	180
(Bignoniaceae) *Millingtonia hortensis* L. f.	*Kang Khong*	7	Champasak, Sayabouli, Vientiane	11	Bolikhamxay, Champasak, Luang Prabang, Vientiane	180
(Solanaceae) *Capsicum annuum* L.	*Phik*	36	Luang Prabang, Sayabouli, Vientiane	NA	NA	108
(Rubiaceae) *Hymenodictyon orixense* (Roxb.) Mabb.	*Som Kop*	9	Luang Prabang, Sayabouli, Savanakhet, Vientiane	1	Luang Prabang	80
(Amaryllidaceae) *Allium sativum* L.	*Ka Thiem*	22	Luang Prabang, Sayabouli, Vientiane	NA	NA	66

This table includes the ten most commonly cited plants for treating symptoms of tuberculosis. The plants are listed in order of weighted score. This point system prioritized the plants that are used in many places and documented by many sources. One point resulted from each citation. One point was given for each province where a plant was cited. One point was given for citation in the manuscripts, and one for citation by healer. The three values were then multiplied to give the weighted score in the right column. For example, Elephantopus scaber (Khii Fai Nok Khoum) was listed 18 times (6 PLM + 12 Healer citations) in five different provinces (Bolikhamxay, Champasak, Salavan, Sayabouli, and Vientiane) in two sources (both in the PLM and by healers). $18 \times 5 \times 2 = 180$. Multiplication, rather than addition, allowed the differences to be seen more clearly

All of the plants listed in Table 2.3 were plants named in manuscripts originating from Sayabouli and Vientiane. There weren't any plants named in manuscripts in the westernmost provinces. The westernmost provinces were heavily affected by bombing during the conflict between Vietnam and the United States (Vongsavanh 1984), which may have destroyed manuscripts in these provinces. Most of the interviews took place in Vientiane and Champasak, so it follows that most of the plants reported by healers were in these two provinces. All but two of these plants were named more frequently in the manuscripts than by healers. One possible explanation for this is that there were many more plants cited in the manuscripts than given by healers.

The citations from the PLM give an indication of past plant populations that were distributed widely in a province, thus allowing them to be integrated into local medicines. There are a number of possibilities as to why plants were not mentioned by healers in provinces where they had been cited in the PLM. Healer interviews were limited by time and budgetary constraints. It is possible that this research did not

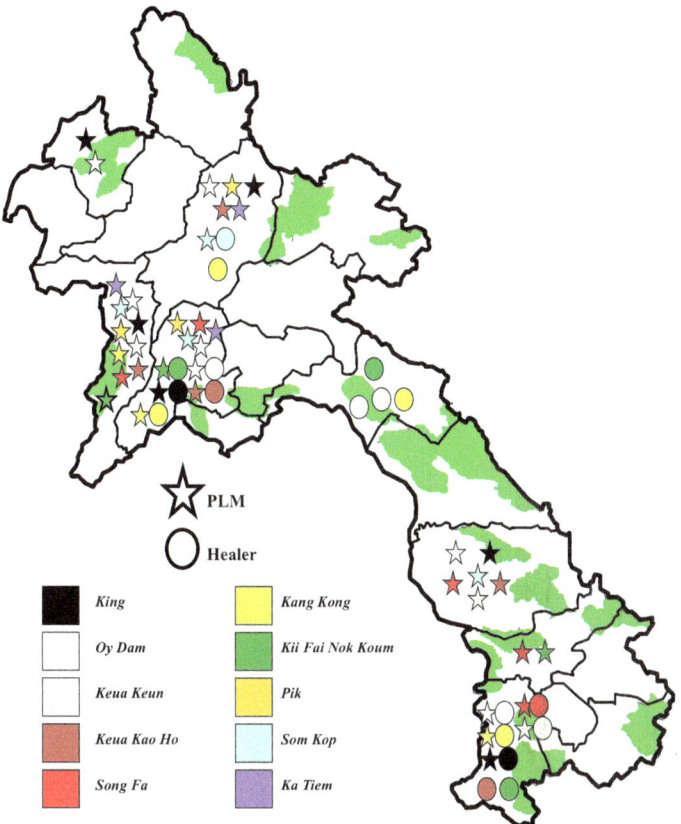

Fig. 2.1 *Report locations of ten most commonly cited plants.* This figure only includes the ten most commonly cited plants to treat respiratory ailments, as listed in Table 2.3. Stars indicate plants listed in the PLM (listed in Table 2.1) and circles indicate plants reported by healers (listed in Table 2.2). The areas shaded in green indicate the National Protected Areas (NPAs). Map adapted from Wikipedia (2012)

reach healers who may have cited the same plants in the corresponding provinces (for example, in Savanakhet). Even when healers were reached and interviews took place, their responses may have been biased. Most of the PLM analyzed, however, were relatively easy to access and were created before this project began, so the information was assumed to be unbiased. It is also possible that the plants were ineffective and have since dropped out of the healers' repertoires. Another possibility, however, is that some plant populations are decreasing due to habitat depletion and/ or overharvesting. This means that plants used fifty or more years ago are no longer easy to access and thus have fallen out of use.

The manuscripts studied came from Champasak, Khammuan, Luang Prabang, Salavan, Savanakhet, Sayabouli, and Vientiane. Healers were interviewed in Bokeo, Bolikhamxay, Champasak, Luang Prabang, and Vientiane. Therefore, no plants

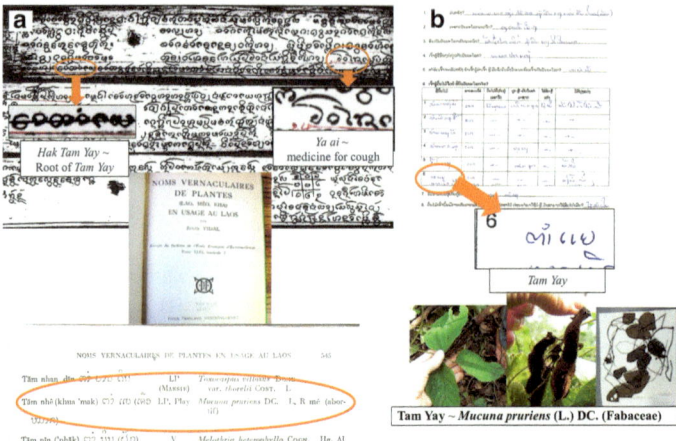

Fig. 2.2 *Example of plant selection, collection, and documentation.* In this example, the plant known as Tam Yay is cited both in the PLM and by a healer. (**a**) When found in the PLM, previous accounts of the common name are used to predict the taxonomic identification. The book image is adapted from Vidal (1959). The digital manuscript image is adapted from the National Library of Laos (NLL 2009). (**b**) When reported by a healer, a herbarium specimen is collected to document the plant. The herbarium specimen is essential for the correct identification of the species

could be cited both in PLM and by healers in Bokeo, Bolikhamxay, Khammuan, Salavan, Savanakhet, or Sayabouli. Most of the reports came from Vientiane, Luang Prabang, and Champasak. These three provinces also hold major cities with high population densities (LSB 2005). Because TB is more of a problem in places where many people live in smaller areas, such as cities, it would make sense that traditional healers in cities would have seen more cases of TB than their rural counterparts. A healer in a city is more likely to see information about TB on billboards or television than a healer in rural Laos, as well.

The total number of plants (1,474) came from 731 different common names. Healers reported 223 of these different common names and the PLM listed 546. It should be noted that there was some overlap, and that some plants were named both by healers and in the PLM. Because there is not a herbarium specimen to support the PLM-derived name for comparison and identification purposes, it is also not known if each of the common names reported by healers or listed in the PLM represents a separate taxonomic species. Figure 2.2 gives an example of one plant, *Tam Yay*, which was cited in the PLM and also documented with a voucher herbarium specimen during an interview with a healer.

Based on the common names, 67 plants named by healers (representing 45 common names and 51 taxonomic species) and 55 plants listed in the PLM (representing 41 common names and 44 taxonomic species) were collected. Some of the plants were collected more than once, accounting for the extra number of the plants collected.

Taxonomic Documentation and Identification

The voucher herbarium specimen is a pressed, dried clipping of an aerial part of a plant with attached leaves, fruits, and/or flowers. The specimen is normally mounted on a specialized, acid-free herbarium sheet of standard 29 by 41.5 cm size. It is a vital piece of evidence for any ethnobotanical work, serving as a reference to the original plant named in an interview or mentioned in a document (Łuczaj 2010). Several duplicates of such specimens from one plant are prepared and deposited in various herbarium institutions for posterity. Herbaria are libraries of plant specimens that contain records of plants. The destruction of herbarium institutions resulting from World Wars I and II in Europe demonstrated the importance of keeping multiple records of a collection on different continents so that if a herbarium were destroyed, a record of the plants deposited in other herbaria would survive.

For this research project, plant collection permission was granted by the Ministry of Agriculture and Forestry of Laos, through the ITM, representing the Ministry of Public Health, under the umbrella of the MOA. Plant samples and their voucher herbarium specimens were collected following the WHO Guidelines on Good Agricultural and Collection Practices for Medicinal Plants (WHO 2003) with attention to the conservation of the species. Standard collection information and field notes were recorded following guidelines described by Alexiades (1996).

Collections of 77 different plant species were made for respiratory ailments, of which 19 species were recollected. A list of all voucher herbarium specimens collected is given in Appendix A, Table A.1. The list includes plants that were collected for other ailments that deserve mention. Two to six duplicate specimens were collected for each plant. Of the specimens collected, one or two duplicates were deposited at the ITM herbarium and one to four were sent to the Field Museum in Chicago, the botanical base of the project. Of the specimens sent to the Field Museum, one set was mounted and deposited at the John G. Searle Herbarium of the Field Museum (F). If available, another set was dispatched for deposit at the Nationaal Herbarium Nederlands (L), a collaborating institution in this project. If other duplicates remained, they were sent to other select herbaria as an exchange for determination or sent to a taxonomic specialist for determination or for confirmation of taxonomic identification, if necessary.

Plant taxonomic identification was carried out by comparison of the taxonomic characteristics of a voucher herbarium specimen with those of a previously identified specimen. Initially, comparisons of the collected specimens with previously identified specimens took place at the ITM herbarium in Vientiane. At a later date, specimen comparison was performed at the Field Museum herbarium (F), as well as with online images of plants and herbarium specimens, especially those posted in the Tropicos® website (http://www.tropicos.org/ImageSearch.aspx), as well as standard floristic treatises, especially Ho (1993), Inthakoun and Delang (2011), and Vidal (1959). Whenever possible, type specimens were accessed from select herbarium institutions.

Chapter 3
Biochemical Validation

This section describes our endeavor to provide biochemical validation and to demonstrate the efficacy of plants used in the treatment of tuberculosis (TB) in the past as recorded in the PLM, and those used for treatment of TB symptoms by contemporary healers, through in vitro laboratory analyses. The research process consists of several steps, starting with the collection of the plant materials, extraction using a standardized chemical solvent, setting up mycobacterial culture, submission of plant extracts to biological assay against virulent *Mycobacterium tuberculosis* H37R*v*, recollection of the active samples for bioactivity-guided fractionation, and eventually, for the elucidation of chemical structures of the active molecules against Mtb.

Methods

Plant Sample Collection and Processing

When a plant was collected for biological testing, two types of material were collected from the same plant: the sample to be tested, and multiple herbarium specimens. The plant sample of the part used in traditional preparations was collected for the actual testing in the laboratory. The herbarium specimens, as described in the Ethnobotany-Taxonomic Documentation and Identification section, were collected to provide evidence or to be a "voucher" for the plant species and are used as a material basis for the taxonomic identification of the species being studied.

Permission for the collection of the plant samples and voucher herbarium specimens was granted by the Government of Laos through a Memorandum of Agreement (MOA) between the University of Illinois at Chicago (UIC) and the Institute of Traditional Medicine (ITM) in Vientiane. For primary screening, 50–250 g of the plant part used by the healer was collected. Samples were selected from clean,

© The Author(s) 2014
B.G. Elkington et al., *Ethnobotany of Tuberculosis in Laos*, SpringerBriefs
in Plant Science, DOI 10.1007/978-3-319-10656-4_3

Fig. 3.1 *Plant sample processing, in preparation for extraction and bioassays.* (**a**) A clean tarp and clean machetes are used in chopping stems into smaller pieces. (**b**) The pieces are allowed to dry in an area protected from insects and animals. Well-ventilated shade-drying and sun-drying methods were used

non-diseased plants, and care was exercised in the collection process in order to minimize the risk of contamination with foreign matter. In collecting leaves, twigs, stems, and branches of woody plants (shrubs, trees), the desired part was cut with a clean machete. For root samples, a small piece was cut some distance from the base of the stem. The lower part of the stem and the root system were left intact, so the plants remained alive to regenerate new roots, stems, and/or branches. Each sample type was placed in a labeled nylon mesh bag and transported to the closest regional TMS, where the sample was chopped into approximately 3 cm cubes or smaller using a clean machete on a clean, plastic tarp. Once chopped, each sample was placed in its original labeled bag. For herbaceous plants, the desired plant parts were put directly into nylon mesh bags to dry. Bagged samples were then dried by placement on a clean concrete platform in a well-ventilated area, according to the protocol designed by Soejarto et al. (2002) (Fig. 3.1b). To decrease the risk of contamination, the area for drying plant samples was kept free from insects, rodents, and birds and was inaccessible to livestock and domestic animals.

In order to test the plants for microbials activity, they needed to be turned into a somewhat homogenous material, achieved through extraction into various solvents. Sample extraction for primary screening was performed at the pharmacognosy laboratories of the ITM in Vientiane. Samples were extracted into 90 % ethanol (EtOH) and repeated twice. The extracts were then condensed using a Heidolph Laborota 4000 rotary evaporator (rotavapor). A small amount (approximately 2.5 g) of each extract was transferred into a plastic or glass container for dispatch to UIC. After arrival at UIC, the extracts were kept in a −20 °C freezer in an effort to minimize compound degradation due to oxidation, exposure to light, heat, or the introduction of contaminants.

Dried plant material that was not extracted at the ITM was sent to UIC and stored (separately) in clean plastic or cloth bags with adequate airflow to prevent the growth of mold at room temperature. All aqueous extracts were filtered with 0.22 μm filters prior to the bioassays to ensure sterility.

Primary Evaluation Protocol

All of the biological assays were conducted in the laboratories of the Institute for Tuberculosis Research (ITR) at UIC. In the primary screening, the purpose was to determine if an extract specifically inhibited virulent *Mycobacterium tuberculosis* H37Rv (Mtb), or if the extract was a general cytotoxin. This primary screening was conducted against *Staphylococcus aureus*, *Escherichia coli*, *Candida albicans*, *Mycobacterium smegmatis*, and Mtb.

The primary screening also included testing for potential toxicity to human cells. In vitro biological testing using Vero cells can be used to predict mammalian cytotoxicity (Cantrell et al. 1996). The Vero lineage of cells originates from 1962 when cells were isolated from the kidney of an African green monkey. Since then, the cells have been grown in culture under controlled conditions. The testing protocol used in this research followed the methods used by Falzari et al. (2005). The results indicate the amount of extract that inhibits growth or metabolism of the culture by 50 %. If the highest concentration of extract failed to inhibit cell growth by 50 %, the percent inhibition is indicated in parentheses. For example, >100 μg/mL (39 %) means that growth was inhibited by 39 % at 100 μg/mL.

Testing against mycobacteria specifically (*M. smegmatis* and virulent Mtb) entailed the use of the microplate Alamar Blue assay (MABA), a relatively fast and inexpensive assay. The reduction of Alamar Blue reagent to a pink color by living cells is an easy way to visually check for activity, which can also be measured fluorometrically with a relatively high level of sensitivity (Collins and Franzblau 1997; Franzblau et al. 1998). In this case, the color and fluorescence in the test wells is compared to the color and fluorescence of the control wells, where the fluorescence is indicative of bacterial growth. A value of 100 % inhibition indicates that there was no net growth. Because the percent inhibition reading is based on an average of the wells containing no mycobacteria, it is possible to have values of more than 100 % inhibition. Similarly, if the net fluorescence is higher than the control wells containing only bacteria, it is possible to get inhibition values of less than 0 %. If an extract exhibited greater than 90 % inhibition of Mtb growth, then the Minimum Inhibitory Concentration (MIC), or the smallest concentration of the extract required to inhibit 90 % of the Mtb growth, was sought.

In order to determine if the active components possibly target non-replicating persistent *Mycobacterium tuberculosis* (NR Mtb), this research utilized the Low-Oxygen-Recovery Assay (LORA) (Cho et al. 2007), which was developed in the Institute of Tuberculosis Research at UIC. Most standard assays for high throughput screening detect activity against rapidly growing bacteria, but the LORA uses mycobacteria that have been deprived of oxygen and are non-replicating.

If a sample showed promising activity, signaled by a high percent inhibition of Mtb, it was recollected. After the primary collection and taxonomic identification, a literature search was conducted to ensure that recollection would not pose a threat to the species abundance. A search for species abundance was conducted online utilizing the CITES list (http://www.cites.org/eng/app/appendices.php) and the Red

List of Threatened Species™ (http://www.iucnredlist.org/). In addition, a literature search for previous research involving *Mycobacteria* was conducted with NAPRALERT® (http://www.napralert.org), PubMed (http://www.ncbi.nlm.nih.gov/pubmed), Embase (http://www.embase.com/), and Scifinder® (http://www.cas.org/products/scifindr/sfweb/).

The results of the initial primary screening results are presented in a paper by Elkington et al. (2014b).

Results of Biological Evaluations

All Species Evaluated

Seventy-seven (77) species acquired through ethnobotanical studies were collected and submitted to biological evaluation. From the 77 total species evaluated, 12 exhibited above 90 % inhibition against *Mycobacterium tuberculosis* H37R*v* (Mtb) at 100 μg/mL in the first evaluation. MIC values from these plants ranged from 0.05 to 96.6 μg/mL, as shown in Table 3.1 and previously published by Elkington et al.

Table 3.1 Crude plant extracts exhibiting Greater Than 90 % Inhibition of Mtb

Scientific name (Collection number)	Common name	MIC (μg/mL) Mtb
(Annonaceae) *Marsypopetalum modestum* (Pierre) B. Xue & R.M.K. Saunders (bge080, 113, 115, 253)	*Tin Tang Tia/Tin Tang Teeyah*	0.05–11.9
(Annonaceae) *Rollinia mucosa* (Jacq.) Baill. (bge104)	*Khantaloht*	43.9–75.2
(Annonaceae) *Uvaria* cf. *microcarpa* Champ. ex Benth. (bge241)	*Phii Phouan*	43.2 to >100
(Annonaceae) *Uvaria rufa* Blume (bge137)	*Tin Tang Tia*	33.1 to >100
(Bignoniaceae) *Fernandoa* cf. *adenophylla* (Wall. ex G. Don) Steenis (bge051, 117)[a]	*Kae Pa*	79.7 to >100
(Menispermaceae) *Tinospora crispa* (L.) Hook. f. & Thomson (bge122, 240, 244)	*Kheua Khao Hor*	2.43–96.2
(Rutaceae) *Aegle marmelos* (L.) Corrêa (bge093, 118)[a]	*Mak Toum*	47.8 to >100
(Rutaceae) *Clausena harmandiana* (Pierre) Guillaumin (bge256)	*Song Fa*	83.1 to >100
(Rutaceae) *Feroniella lucida* Swingle (bge066, 110)[a]	*Kok Sung*	90.4 to >100
(Rutaceae) *Glycosmis pentaphylla* (Retz.) DC. (bge248)	*Som Xeun*	93.5 to >100
(Rutaceae) *Micromelum minutum* Wight & Arn. (bge83, 114)[a]	*Sa Mat Khao*	45.7 to >100
(Verbenaceae) *Vitex trifolia* L. (bge087, 243)[a]	*Phii Seua*	77.6 to >100

Entries in this table are listed in alphabetical order by the taxonomic identity of each species. The collection numbers in parentheses represent the voucher herbarium specimen numbers. All extracts were tested at 100 μg/mL
[a]Recollections of species marked by an asterisk did not exhibit activity in the bioassays

(2009), where the spelling for the common name is *Tin Tang Teeyah* and the taxonomic identity was *Polyalthia* sp., and later, when the spelling for the common name was changed to *Tin Tang Tia* and the taxonomic identity was determined to be a *Marsypopetalum* species (Elkington et al. 2014b).

Case Study #1: Commercial Remedies

Traditional remedies often contain many plants mixed together. In order to test some of these combinations of plants that would be given to patients in a traditional setting, recommendations were sought from a traditional medicine vendor at the ITM's government-sponsored clinic and from outside traditional medicine vendors. The vendors were informed about the goals of this research project, to which they graciously complied to help. When asked about remedies to be given to a person with symptoms of TB, one ITM vendor recommended a treatment for "lung infection" (*aksep poht*). On the other hand, another traditional healer recommended a treatment for "deformed lungs" (*poht phiikan*). Each of the two compound remedies consisted of a combination of three plant species. In collaboration with the ITM, samples of each plant were collected for separate biological evaluation.

The traditional remedy for "lung infection" (*aksep poht*-TR1) contained a mixture of three different plant species, including aerial parts of *Lygodium* sp. (Lygodiaceae) (*Koot Ngong*), small pieces of the stem of *Irvingia malayana* Oliver ex Bennett (Simaroubaceae) (*Bohk*), and small pieces of the stem of *Mitragyna* sp. (Rubiaceae) (*Thohm Phai*). See Fig. 3.2.

In order to test the potential safety and effectiveness of this whole (multi-component) traditional remedy, a decoction of one package of dried plant materials was prepared according to the instruction inserted in the package. This entailed boiling the contents of the package (approximately 30 g of dried plant materials) in

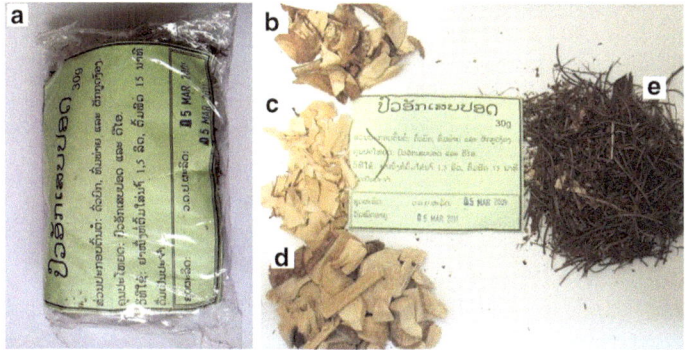

Fig. 3.2 *Traditional remedy 1.* (**a**) Packaged remedy. (**b**) *Thohm Phai* stem. (**c**) *Bohk* stemwood. (**d**) *Bohk* stem. (**e**) *Koot Ngong* aerial parts

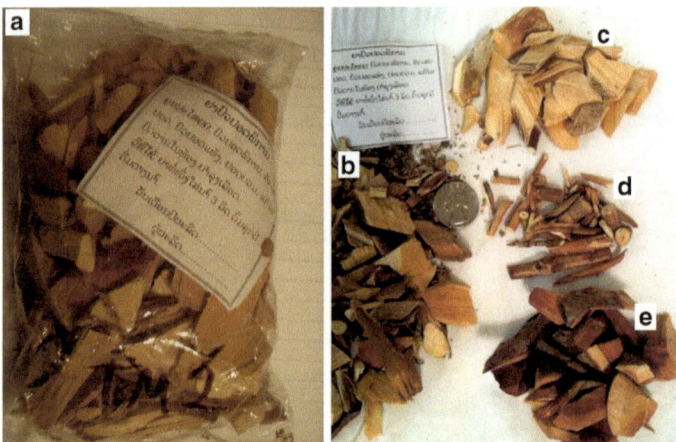

Fig. 3.3 Traditional remedy 2. (**a**) Packaged remedy. (**b**) Mixture of three plants. (**c**) *Khom La Wan Joh* stem. (**d**) *Hing Hai* aerial parts. (**e**) *Som Xeun* dried stem

1.5 L of water for approximately 15 min. For the bioassays, a 5 mL aliquot at a concentration of 7.2 mg/mL was taken and filtered to ensure sterility. A small part of the extract was submitted directly to the bioassay as an aqueous extract and another small part was dried and redissolved in dimethyl sulfoxide (DMSO) to make a concentration of 10 mg/mL.

The traditional remedy for *poht phiikan*, ("deformed lungs"-TR2) consisted of a combination of dried aerial parts of *Crotalaria pallida* Aiton (Fabaceae) (*Hing Hai*), dried and milled stem of *Glycosmis pentaphylla* (Retz.) DC. (Rutaceae) (*Som Xeun*), and dried and milled stem of *Melicope* cf. *pteleifolia* (Champ. ex Benth.) T.G. Hartley (Rutaceae) (*Khom La Wan Joh*). See Fig. 3.3.

The water and DMSO extracts of this commercial remedy were prepared in a similar manner as used for the first traditional remedy. One package of dried plant materials (approximately 75 g) was boiled in 1.5 L of water for approximately 20 min. A small amount (5 mL) was filtered and submitted directly to bioassays as an aqueous extract at a concentration of 8.0 mg/mL. Another small amount was dried and redissolved in DMSO at a concentration of 10 mg/mL.

In both cases, for both the combination of plants as well as the individual plants, the results showed negligible in vitro inhibition of Mtb growth, as well as negligible cytotoxicity against Vero cells.

Case Study #2: Kheua Khao Hor

Analysis of the lists of plants in the Ethnobotany section from the interviews and palm leaf manuscripts indicated that this plant has been widely used to treat respiratory ailments for many years. *Kheua Khao Hor* was cited by two healers in

Fig. 3.4 *Kheua Khao Hor*. (**a**) Photo of stem and leaves. (**b**) Commercial product purchased from Thai grocery store. (**c**) Herbarium specimen of bge240. (**d**) Herbarium specimen of bge244

Champasak and Vientiane, as well as 17 times in the PLM, ranging from Luang Prabang, Sayabouli, Savanakhet, and Vientiane, as shown in Table 2.3. After *Kheua Khao Hor* (*Tinospora crispa* (L.) Hook. f. & Thomson, Menispermaceae) showed significant activity in the MABA against Mtb growth, it was selected for further fractionation and analysis.

This vine is native to Southeast Asia, growing by twining on trees to more than 20 m above ground, with a stem diameter 1–3 cm, and dotted with many blunt tubercles (Fig. 3.4a). The leaves are alternate and cordate in shape. Although not seen in these collections, the flower is yellow-green in color, and the fruit, a drupe, is orange in color. A more complete taxonomic description of this species is provided in the *Flora of China* (1994b), and additional information can be found from IPNI (2008), Tropicos® (2011), and NAPRALERT® (2010).

Though the plant is found abundantly throughout Southeast Asia, no previous literature on the testing of this plant against Mtb could be found. It was first collected in April 2009 (bge122) and recollected twice in August 2009 (bge240 and bge244). Additionally, a commercial sample under the name of "Thai Galanga" was purchased in a Thai grocery store in Chicago in 2011. The shopkeeper asserted that the contents were *Tinospora crispa*, even though the name "Thai Galanga," a Zingiberaceae species, was printed on the outside of the bag. Macroscopic identification revealed that the contents inside of the bag looked identical to the plant samples of *Kheua Khao Hor* collected in Laos (Fig. 3.4b).

In the end, even though activity reached 90 % inhibition in the early primary assays, retesting did not show the same activity, when the crude extract only reached 4 % inhibition. In retrospect, we feel that additional bioassays should be run on this plant, ideally shortly after collection and drying. Of the two samples that were fractionated, bge240 and the commercial sample, neither showed appreciable activity. If there are active components, it is possible that they degrade over time, or that the components work synergistically and were separated during the fractionation process.

Case Study #3: **Tin Tang Tia**

This plant was reported by a healer as a component in three different formulations consisting of up to 32 different plant species. Traditionally, the stem or root is dried and rubbed on a stone to produce a powder, which is then mixed with water and powder from the other plants and given to the patient to drink. The healer said that it can also be boiled with the other plants, and again consumed as a drink.

Tin Tang Tia was first collected in August 2007 and submitted to the primary bioassays. After confirming that it exhibited the lowest MIC against Mtb of all of the plants collected, recollections of this plant were carried out in both the dry and rainy seasons, and at various stages of reproductive development. Voucher herbarium specimens were prepared for all collections, which were used as basis for taxonomic identification and for biological evaluation. See Fig. 3.5 for photos of the plant.

In the process of searching for this plant in the field based on its common name, *Tin Tang Tia*, as the term of inquiry, confusion ensued. When asked to identify *Tin*

Fig. 3.5 *Tin Tang Tia.* The fruit and flowers are both shown in this picture

Fig. 3.6 Species referred to as *Tin Tang Tia*. (**a**) (Annonaceae) *Marsypopetalum modestum* (Pierre) B. Xue & R.M.K. Saunders. (photo of bge115 and herbarium specimen of bge080). (**b**) (Annonaceae) *Anomianthus dulcis* (Dunal) James Sinclair (bge124). (**c**) (Annonaceae) *Uvaria rufa* Blume (bge137)

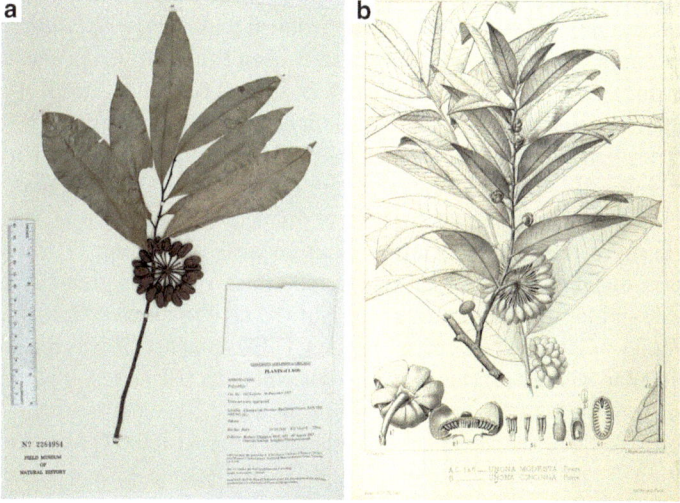

Fig. 3.7 First collection of *Tin Tang Tia* and the first published image of the species. (**a**) Collection bge080 before final identification. (**b**) Adapted from the sketch in the original description of *Unona modesta* (Pierre 1881)

Tang Tia, healers referred to multiple different taxonomic species. Photographs of three of these species, *Marsypopetalum modestum*, *Anomianthus dulcis*, and *Uvaria rufa*, all members of the Annonaceae family, are presented in Fig. 3.6. To add to the puzzle, when shown a photo of bge080 and asked for the common name, other healers often gave the common name of *Phii Phouan*. *Uvaria rufa* was collected twice, first for the stomach and nervous system and identified as *Mak Phii Phouan* (bge132), and second to treat a cough and identified as *Tin Tang Tia* (bge137).

After consulting many different texts and taxonomic specialists, in addition to DNA sequencing, the species was determined to be *Marsypopetalum modestum* (Pierre) B. Xue and R.M.K. Saunders. The original description of this species made by Pierre (1881) is in Fig. 3.7, along with the first collection for this research, which matches the characteristics of bge080, 113, 115, 253, and 255. Based on further

Table 3.2 *Tin Tang Tia* bioassay data

Collection number	Plant part	Extraction solvent	MIC (µg/mL)		IC$_{50}$ (µg/mL)	SI	
			Mtb	NR Mtb	Vero	Mtb	NR Mtb
bge080	Stem	EtOH	1.33	5.85	51.49	38.71	8.80
bge113	Stem	EtOH	13.93	22.87	60.95	4.38	2.66
bge115	Stem + root	EtOH	6.58	8.50	47.52	7.22	5.59
bge253	Stem	EtOH	5.97	2.50	<6.25	<1.0	<2.5
bge253	Stem	H$_2$O	23.52	5.95	>100 (81 %)	>4.25	>16.81

The collection number represents the voucher herbarium specimen number. Extracts were tested at 100 µg/mL

taxonomic research by taxonomists, the species was transferred from the genus *Unona* to the genus *Polyalthia* and is now grouped with *Marsypopetalum*.

M. modestum are small trees, which have been found in peninsular Southeast Asia, growing 3–4 m in height. The trunk is around 3.5 cm in width. Leaves are simple, alternate, acuminate, 3–8 cm in width by 8–25 cm in length, on 0.5 cm long petioles. They bear the characteristics of other *Marsypopetalum* leaves, with straight secondary veins and prominent arcuate loops (Xue et al. 2011), but immersed reticulation. Flowers are approximately 1 cm in diameter, with numerous stamens and greenish fleshy petals. The apocarpous fruit consists of a group of fruitlets in an umbelliform manner, as extra-axillary clusters. Each fruitlet is glabrous, ellipsoid, one-seeded, and turning from green to bright red as it matures. While this species was not found in the IUCN Red List or the CITES database, it was not frequently encountered and deserves further study to determine its risk status. It was only seen in two locations in all of the field expeditions in Laos.

In order to test this plant, ethanol extracts were created at the ITM's laboratories. The resulting extracts were redissolved in DMSO for testing. An aqueous extraction was also performed at UIC, where approximately 5 g of dried stem were boiled in 200 mL water.

The results of the primary biological evaluation of the extracts bge080 through bge253 are presented in Table 3.2.

The values for cytotoxicity, listed in the IC$_{50}$ column in Table 3.2, are calculated and compared to the MIC values through calculation of a Selectivity Index (SI) for each extract through the following formula: $SI = IC_{50}/MIC$, seen in the far right columns. A higher value indicates a higher degree of selectivity to Mtb than to mammalian cells.

Based on the high SI value for selectivity for Mtb, bge080 was fractionated and the fractions were resubmitted to the bioassays. The next highest SI value from Table 3.2 was from bge115, which contained both stem and root material. In both cases, the crude EtOH extracts were fractionated with a SPE cartridge to give six fractions. Fractionation with reversed-phase HPLC was then used to purify active isolates. The structure of bge080 isolate D and bge115 isolate 3 were then elucidated as 2,2′-dithiobis(pyridine *N*-oxide), known more commonly as dipyrithione (Fig. 3.8).

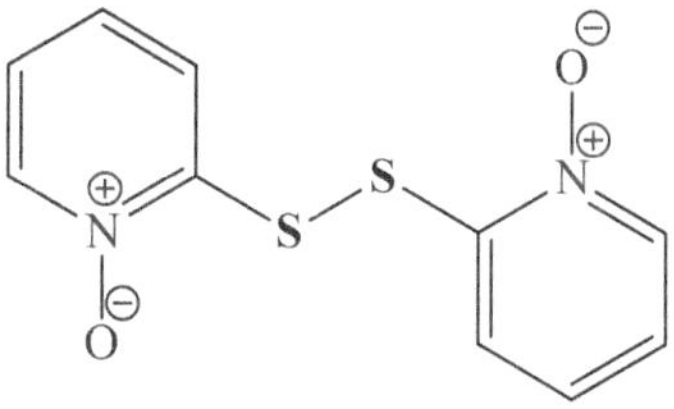

Fig. 3.8 *Dipyrithione*. Known more commonly as dipyrithione, 2,2′-dithiobis(pyridine *N*-oxide), was isolated from *Marsypopetalum modestum* (Pierre) B. Xue & R.M.K. Saunders

Dipyrithione is commercially available and is currently used as a pesticide and fungicide. Alternate names are di-2-pyridyl disulfide, 1,1′-dioxide (IUPAC). Other generic names include omadine disulfide (IS: Olin) and OMDS (IS: Olin). It is currently used in the anti-dandruff shampoos, Crimanex® (Drossapharm, Switzerland) and Perkapil® (Kurtsan, Turkey). The Chemical Abstracts Service (CAS) registry number is 3696-28-4 (Drugs.com 2012). It has also been shown to inhibit inflammation (Han et al. 2010; Huang et al. 2011) and to reduce lung inflammation and some lung injury in mice (Huang et al. 2011).

While dipyrithione does have pesticidal activity, its presence in *Marsypopetalum modestum* does not seem to result from contamination with agricultural pesticides. It has been isolated from other natural products (Nicholas et al. 2001; O'Donnell et al. 2009), and similar compounds have been reported from the closely related species, *Trivalvaria costata* (Hook. f. & Thomson) I.M. Turner (Lu et al. 2010). In addition, other plants that were collected from the same area as bge080 (bge079 through bge084, bge114) did not exhibit the same activity in the assays. This species was also collected from two different locations, one of which was in the wild, and both collections exhibited similar activity and contained this compound, as demonstrated by NMR and LC-MS (Elkington 2013).

Chapter 4
Summary and Conclusions

Summary of the Findings

The goals of this project were (1) to document the traditional medicinal uses of plants of Laos, with specific focus on plants used to treat symptoms of tuberculosis (TB), and (2) to provide biochemical validation of the claimed use. Data on the use of plants to treat symptoms of TB were explored from old medical documents in the form of palm-leaf manuscripts, as well as acquired through field interviews with contemporary healers. The collected plants were submitted to biological assays against virulent *Mycobacterium tuberculosis* H37Rv (Mtb), followed by the isolation and elucidation of active molecules.

In order to determine which plants have been in use over a long period of time, centuries-old medical documents from seven provinces in Laos were analyzed. Through this process, entries for 315 treatments for symptoms related to tuberculosis were translated into the modern Lao and English languages. Further analysis of these entries gave a list of 1,133 plants that have been used to treat symptoms associated with tuberculosis. Follow-up research involved field interviews with 58 contemporary healers in five different provinces about plants that they are currently using for disease treatment, which resulted in an additional list of 341 plants.

Fieldwork in Laos was carried out from 2007 to 2010. The medical manuscript sources came from seven provinces, including Champasak, Khammuan, Luang Prabang, Salavan, Savanakhet, Sayabouli, and Vientiane, ranging from all regions of the country, but excluding westernmost provinces. Four written scripts used to transcribe four written languages were found, with the majority in Tham-Lao. While most of the manuscripts did not indicate the era in which they were created, based on similarities in terminology and script, it was concluded that most were created in the last 200 years. Complete entries for 315 treatments for symptoms of tuberculosis were translated into the modern Lao and English languages. A catalog of 7,080 ailments was prepared to serve as a reference for future research. The entries describing respiratory ailments gave a list of 1,133 plants used to treat symptoms

© The Author(s) 2014
B.G. Elkington et al., *Ethnobotany of Tuberculosis in Laos*, SpringerBriefs
in Plant Science, DOI 10.1007/978-3-319-10656-4_4

associated with tuberculosis. Treatments for "fever" were frequent, as was expected due to the many illnesses with fever as a main symptom.

Field interviews with 58 healers took place in five provinces: Bokeo, Bolikhamxay, Champasak, Luang Prabang, and Vientiane. The majority of interviews were performed in Champasak and Vientiane provinces. Most of the healers were male between the ages of 30 and 80. They represented ten different ethnic groups, and three major religious groups. Professionally, they were civil servants, farmers, and/ or Buddhist monks/nuns. Perceptions about TB were mainly congruent with biomedical viewpoints in that the majority of the healers indicated that TB had to do with the lungs, and specifically with the main symptom of coughing, and that it was communicable by being around somebody who had it. Most healers (26 out of 34) said they would turn to traditional medicine first if they suspected they or somebody in their family had TB.

In regard to other diseases, most of the healers stated that they treat stomach ailments most frequently. Each healer favored unique plants, a phenomenon which may have been due to the different ecosystems in which each healer lived. Nevertheless, *Cyclea barbata* Miers (*Mor Noy*) and *Tiliacora triandra* (Colebr.) Diels (*Yanang*), both belonging to the same family, Menispermaceae, came up most frequently as treatments for a wide variety of ailments. Both of these plants are found widely throughout Laos and Southeast Asia. There was a wide mix of where the healers said they had gained their knowledge of plants. Most claimed to have been taught by family members and/or other healers, but nine claimed to have learned from written documents, including the palm leaf manuscripts. Only one healer claimed to have learned from an independent source (i.e., dreams).

During field interviews, healers would often point out plants they use to treat respiratory ailments. Samples and voucher herbarium specimens of these plants would be collected, along with plants recognized by the researchers that were listed in the PLM. Of the specimens collected, one or two were deposited at the ITM herbarium and one to four were sent to the Field Museum in Chicago for deposit at this institution and for distribution to other institutions. Of the specimens sent to the Field Museum, one was deposited at the Field Museum herbarium (F), the botanical base of this research. If available, another was deposited at the Nationaal Herbarium Nederlands (L), a collaborating institution. Any additional specimens were dispatched to other herbaria or taxonomic specialists for determination or confirmation of taxonomic identity, as necessary. A total of 94 plants were collected and deposited at the ITM herbarium in Vientiane and the Field Museum Herbarium in Chicago (F), with an additional 57 deposited at the National Herbarium at Leiden (L). Of these, 43 plants were listed in the manuscripts and 50 were named by healers. The six plants contained in the commercial remedies were a part of these collections.

A major constraint for plant collection was inconsistency of the common names. While a healer could point out a plant in question, plants recorded in the manuscripts were determined solely based on their common names. Many plants known under a single common name have represented multiple taxonomic species. In cases where one common name referred to more than one plant, efforts were made to collect all of the plants.

Samples of the plant part traditionally used were collected for 44 plants listed in the manuscripts and 54 named by healers (20 were named in both the manuscripts and by the healers). In addition, this research also undertook the testing of multi-component traditional remedies and analyzed two commercial products. Each of these two formulations contained three plants, for which samples were also collected for separate testing.

Plant taxonomic identification was carried out by comparing voucher herbarium specimens to a previously identified specimen in deposit in an herbarium, as well as with taxonomic circumscriptions and illustrations in standard floristic treatises (Ho 1993; Inthakoun and Delang 2011; Vidal 1959). Identification efforts began at the ITM herbarium, and later continued at the Field Museum herbarium (F), along with the use of the online database and images retrievable from the Tropicos® database (http://www.tropicos.org/ImageSearch.aspx), along with other databases of the world's renowned herbaria. Special effort was made to find type specimens posted online. In the end, taxonomic specialists were consulted for select species confirmation.

Investigation of the in vitro safety and efficacy of different traditional remedies from Laos to treat tuberculosis (TB) was executed by observing whether specific plants have been used consistently through time to treat the symptoms of TB. The idea was that if specific plants are effective and not harmful, people will continue to use those plants. While consistent plant usage through time serves as proof of the plants' medicinal safety, biological evaluation followed by the fractionation and isolation of active components and subsequent cytotoxicity assays in this research have helped to affirm the degree of safety and efficacy of the plants investigated. All of the plant samples were extracted into methanol, ethanol, and/or water and submitted to various biological assays in vitro in order to determine the amount of growth inhibition of virulent *Mycobacterium tuberculosis* H37R*v* (Mtb), as opposed to other microbes and mammalian Vero cells. Of all of the collected species, 10 % of the plants named by healers (8 of 77), 9 % named in the PLM (7 of 77), and 4 % named by both (3 of 77) were active (defined as exhibiting greater than 90 % inhibition in the MABA or LORA). This gave a total of 15.6 % plants exhibiting anti-Mtb properties (12 active of 77 plants). However, this research only looked at in vitro inhibition of Mtb and Vero cells. The plants and formulations studied may well have other healing properties for respiratory ailments that were beyond the scope of this research, such as analgesic, antitussive, or immune system boosters, but these properties were beyond the scope of this research.

To test plants in combinations, two traditional remedies were purchased. One was to treat "lung infection" (*aksep poht*—TR1) and another to treat "deformed lungs" (*poht phiikan*—TR2). Additional collections of each of the six plant components of these two treatments were made for individual testing. While TR1 and TR2 did not appear to be cytotoxic in vitro, both exhibited negligible activity against Mtb. Separating out the individual plants and extracting with methanol did not elicit a significant change in percent inhibition or cytotoxicity, except in the case of *Crotalaria pallida*, which demonstrated inconsistent activity. Two different species of *Lygodium* going by the same common name (*Koot Ngong*—reported by four healers) were evaluated and showed no appreciable in vitro activity in either assay. While *Irvingia malayana* (*Bohk*) was cited by eight different healers in three different provinces, it was not cited in the PLM. It is possible that the name given by the healers was somehow altered from names used in the recent past, which would appear in the manuscripts. Though it was an ingredient of TR1 and reported by one healer, *Thohm Phai* (*Mitragyna* sp.) was not found in the PLM. Despite the negligible inhibition of Mtb seen in this research, previous literature was found indicating analgesic and antitussive properties common in the genus *Mitragyna* (Adkins et al. 2011; Barger et al. 1939; Kang et al. 2006). *Crotalaria pallida* (*Hing Hai*) was named by three healers and once in the PLM. Although it exhibited activity in one bioassay, the same activity was not seen in recollections. *Glycosmis pentaphylla* was cited by three healers and twice in the PLM. It was interesting that *Khom La Wan Joh* (*Melicope* cf. *pteleifolia*) was cited by four healers, but not cited in the manuscripts. The bioassays for *Khom La Wan Joh* also showed negligible activity against Mtb.

Bioassay-guided fractionation was carried out on two of the plants that had shown high inhibition of Mtb, namely, *Tinospora crispa* (L.) Hook. f. & Thomson (Menispermaceae) and *Marsypopetalum modestum* (Pierre) B. Xue & R.M.K. Saunders (Annonaceae). The fractions from *Tinospora crispa* exhibited inconsistent inhibition of Mtb. Bioassay-guided fractionation of *Marsypopetalum modestum* led to the isolation and elucidation of anti-mycobacterial compound dipyrithione, isolated for the first time from the genus *Marsypopetalum* through this research.

Conclusions

This research has provided documentation about medicinal plants of Laos, with special reference to plants used to treat symptoms of TB. This document should help to increase awareness of Laos' rich diversity of medicinal plants. Hopefully, it will also provide incentive for the preservation of the undeveloped forested areas that remain, which still hold a wealth of medical information for future discoveries. A great deal of the traditional healing art in Laos is well-preserved in historical palm leaf manuscripts and in knowledge held by traditional healers. With the help of contemporary healers, additional analysis of the medical manuscripts could lead to invaluable new discoveries about alternative medical treatments and the medical history of the people of Laos. In-depth studies by the scholarly community of the manuscripts and knowledge held by healers help to demonstrate their importance, while affirming the indigenous intellectual property of disease treatments of the people of Laos, further encouraging their preservation. This research touched on many aspects of traditional medicine use and serves as justification and a starting point for further research.

This research has also achieved the second goal, namely, to provide scientific support and biochemical validation of the safety and efficacy of different traditional remedies from Laos to treat TB through in vitro biological assays. This was attempted by evaluating plants that have been used consistently for many years to treat the symptoms of TB. While proof of consistent plant usage through time serves as a demonstration of the plants' medicinal safety, the fractionation and isolation of active components and subsequent cytotoxicity assays has helped to affirm the degree of safety of the active components.

In Roszak's book on ecopsychology, or connecting psychology and ecology with political and practical goals, he asserts that, "For us the way *back* in time is not the way *out* of our environmental crisis. If any part of an Animist sensibility is to be reclaimed, the project will have to integrate with modern science" (1993). While not all healing systems and techniques are translatable through hard scientific terms, this research has taken on the goal to encourage retention and passing of medicinal plant traditions from one generation of healers to the next through the translation of some traditional treatments into biomedical terms. It is hoped that this and similar types of research will increase awareness of Laos' rich medicinal plants and plant diversity and provide incentive to conserve the undeveloped forested areas that remain, which still hold a wealth of information for future discoveries.

All of the plants encountered through this research hold doubtless healing properties, regardless of the mycobacteria bioassay values. If there is no inhibition of Mtb, perhaps they aid in boosting the immune system, relieving aches and pains, which were beyond the scope of this research. For example, *Mitragyna* species (*Thohm Phai*) have demonstrated both analgesic and antitussive properties (Adkins et al. 2011; Takayama 2004), and *Justicia adhatoda* (*Hou Ha*) contains powerful anti-inflammatory and anesthetic components (Grange and Snell 1996; Gupta 2010).

While taking a plant from its native surroundings and being able to truly recreate the traditional remedy and the accompanying circumstances in a laboratory may not be possible at this point in time, this project has biochemically demonstrated the potential healing power of some traditionally used plants from Laos.

Appendix A
Additional Tables

Table A.1 Plants collected for this research

Voucher herbarium specimen number (Family) Latin binomial	Common name	Field museum (F) Accession number	Use	Notes
bge043 (Asteraceae) *Elephantopus scaber* L.	*Khii Fai Nok Khoum*	2284958	Respiratory	Reported by 12 healers and 6 PLM entries
bge044 (Rubiaceae) *Benkara depauperata* (Drake) Ridsdale	*Kheua Khat Khao*	2284959	Respiratory	Reported by 2 PLM entries
bge045 (Rhamnaceae) *Colubrina javanica* Miq.	*Khan Toum*	2284960	Respiratory	Reported by 1 PLM entry
bge046 (Bignoniaceae) *Millingtonia hortensis* L. f.	*Kang Khong*	2284961	Respiratory	Reported by 11 healers and 7 PLM entries
bge047 (Burseraceae) *Canarium* cf. *hirsutum* Willd.	*Kok Keuam*	2285011	Respiratory	Reported by 1 healer
bge048 (Araliaceae) *Heteropanax fragrans* Seem.	*Oy Xang*	2285084	Respiratory	Reported by 3 healers and 3 PLM entries
bge049 (Araliaceae) *Schefflera* sp.	*Tin Nohk*	2285017	Respiratory	Reported by 1 healer
bge050 (Bignoniaceae) *Oroxylum indicum* (L.) Kurz	*Lin Mai (mak)*	2285012	Respiratory	Reported by 3 healers and 1 PLM entry
bge051 (Bignoniaceae) *Fernandoa* cf. *adenophylla* (Wall. ex G. Don) Steenis	*Kae Pa*	2285013	Respiratory	Reported by 2 healers

(continued)

© The Author(s) 2014
B.G. Elkington et al., *Ethnobotany of Tuberculosis in Laos*, SpringerBriefs in Plant Science, DOI 10.1007/978-3-319-10656-4

Table A.1 (continued)

Voucher herbarium specimen number (Family) Latin binomial	Common name	Field museum (F) Accession number	Use	Notes
bge052 (Celastraceae) *Salacia chinensis* L.	*Tah Kai*	2284971	Respiratory	Reported by 2 healers
bge053 (Stemonaceae) *Stemona cochinchinensis* Gagnep.	*Sam Sip (hua)*	2285016	Respiratory	Reported by 1 healer, 1 PLM entry
bge054 (Arecaceae) *Arenga caudata* (Lour.) H.E. Moore	*Tao Hang*	2285014	Respiratory	Reported by 1 healer, 3 PLM entries
bge055 (Fabaceae) *Millettia* sp.	*Hang Yen*	2285015	Respiratory	Reported by 1 healer
bge056 (Moraceae) *Ficus erecta* Thunb.	*Deua Pong*	2284970	Respiratory	Reported by 1 healer
bge057 (Lygodiaceae) *Lygodium microphyllum* (L.) Sw.	*Koot Ngong*	2284969	Respiratory	Reported by 4 healers
bge058 (Rubiaceae) *Mitragyna rotundifolia* (Roxb.) Kuntze	*Thohm Phai*	2284968	Respiratory	Reported by 1 healer
bge059 (Araceae) *Lasia spinosa* (L.) Thwaites	*Bo Nam/Phak Nam*	2285018	Respiratory	Reported by 1 healer
bge060 (Rubiaceae) *Psychotria* sp.	*Kuk Mohk*	2284967	Respiratory	Reported by 1 healer
bge061 (Rutaceae) *Melicope pteleifolia* (Champ. ex Benth.) T.G. Hartley	*Khom La Wan Joh*	2284966	Respiratory	Reported by 4 healers
bge062 (Lygodiaceae) *Lygodium flexuosum* (L.) Sw.	*Koot Ngong/Koot Khee Pa*	2284965	Respiratory	Reported by 4 healers
bge063 (Rubiaceae) *Ixora* sp.	*Khai Nao (noy)*	2284964	Respiratory	Reported by 1 healer
bge064 (Irvingaceae) *Irvingia malayana* Oliver ex Bennett	*Bohk*	2285019	Respiratory	Reported by 8 healers
bge065 (Solanaceae) *Solanum melongena* L.	*Mak Kheua Kheun (hak)*	2284963	Respiratory	Reported by 11 healers and 9 PLM entries
bge066 (Rutaceae) *Feroniella lucida* Teijsm. & Binn.	*Kasung (mak/kok)*	2285001	Respiratory	Reported by 1 PLM entry

(continued)

Table A.1 (continued)

Voucher herbarium specimen number (Family) Latin binomial	Common name	Field museum (F) Accession number	Use	Notes
bge067 (Chrysobalanaceae) *Parinari anamensis* Hance	*Phohk*	2284962	Respiratory	Reported by 1 healer
bge068 (Acanthaceae) *Justicia* cf. *adhatoda* L.	*Hou Ha* (*kok*)	2284972	Respiratory	Reported by 1 healer, 6 PLM entries
bge069 (Meliaceae) *Sandoricum koetjape* (Burm. f.) Merr.	*Khor Phou*	2284973	Respiratory	Reported by 1 healer
bge070 (Tiliaceae) *Microcos paniculata* L.	*Khom Som*	2284974	Respiratory	Reported by 1 healer
bge071 (Melastomataceae) *Melastoma malabathricum* L.	*Ben Ah/En Ah*	2284975	Respiratory	Reported by 1 healer
bge072 (Lauraceae) *Litsea cubeba* (Lour.) Pers.	*Sii Khai Tohn*	2284976	Respiratory	Reported by 5 healers and 1 PLM entry
bge073 (Euphorbiaceae) *Jatropha curcas* L.	*Niao Khao* (*mak*)	2284977	Respiratory	Reported by 1 PLM entry
bge074 (Euphorbiaceae) *Jatropha gossypifolia* L.	*Niao Deng* (*mak*)	2284978	Respiratory	Reported by 1 PLM entry
bge075 (Rhamnaceae) *Ziziphus oenopolia* (L.) Mill.	*Nam Lep Meo*	2284979	Respiratory	Reported by 1 PLM entry
bge076 (Euphorbiaceae) *Antidesma bunius* (L.) Spreng.	*Mao* (*mak*)	2284980	Respiratory	Reported by 1 PLM entry
bge077 (Verbenaceae) *Clerodendrum palmatolobatum* Dop	*Phouang Phii Deng*	2284981	Respiratory	Reported by 2 healers
bge078 (Rutaceae) *Micromelum* cf. *falcatum* (Lour.) Tanaka	*Sa Mat*	2284982	Respiratory	Reported by 1 PLM entry
bge079 (Rutaceae) *Glycosmis pentaphylla* (Retz.) DC.	*Som Xeun*	2284983	Respiratory	Reported by 3 healers and 2 PLM entries

(continued)

Table A.1 (continued)

Voucher herbarium specimen number (Family) Latin binomial	Common name	Field museum (F) Accession number	Use	Notes
bge080 (Annonaceae) *Marsypopetalum modestum* (Pierre) B. Xue & R.M.K. Saunders	*Tin Tang Tia*	2284984	Respiratory	Reported by 1 healer
bge081 (Euphorbiaceae) *Phyllanthus emblica* L.	*Oy Sam Souan*	2285002	Respiratory	Reported by 1 PLM entry
bge082 (Myrsinaceae) *Ardisia* sp.	*Tin Cham Kohn*	2284985	Respiratory	Reported by 1 healer
bge083 (Rutaceae) *Micromelum minutum* Wight & Arn.	*Sa Mat Khao*	2284986	Respiratory	Reported by 1 healer
bge084 (Apocynaceae) *Tabernaemontana bufalina* Lour.	*Phet Pa (mak)*	2284987	Respiratory	Reported by 2 PLM entries
bge085 (Loganiaceae) *Strychnos nux-blanda* A.W. Hill	*Toum Kah Khao*	2284988	Respiratory	Reported by 1 healer
bge086 (Sapindaceae) *Dimocarpus longan* Lour.	*Kha Liin*	2284989	Respiratory	Reported by 1 healer
bge087 (Verbenaceae) *Vitex trifolia* L.	*Phii Seua*	2284990	Respiratory	Reported by 7 PLM entries
bge088 (Capparaceae) *Capparis* cf. *micrantha* A. Rich.	*Kheua Khao Mohk*	2285010	Respiratory	Reported by 1 healer, 1 PLM entry
bge089 (Fabaceae) *Cassia tora* L.	*Nya Lap Meun*	2284991	Respiratory	Reported by 4 PLM entries
bge090 (Meliaceae) *Aglaia* sp.	*Phii Mob*	2284992	Respiratory	Reported by 1 PLM entry
bge091 (Capparaceae) *Capparis micrantha* A. Rich.	*Xai Xou Tohn (hak)*	2285085	Respiratory	Reported by 1 PLM entry
bge092 (Euphorbiaceae) *Chaetocarpus castanocarpus* (Roxb.) Thwaites	*Bohk Khai*	2284993	Respiratory	Reported by 4 healers
bge093 (Rutaceae) *Aegle marmelos* (L.) Corrêa	*Mak Toum*	2284994	Respiratory	Reported by 1 PLM entry
bge094 (Fabaceae-Papil) *Mucuna pruriens* (L.) DC.	*Tam Yay*	2284995	Respiratory	Reported by 5 healers and 5 PLM entries

(continued)

Table A.1 (continued)

Voucher herbarium specimen number (Family) Latin binomial	Common name	Field museum (F) Accession number	Use	Notes
bge098 (Fabaceae) *Dalbergia millettii* Benth.	*Padong Khor/Pan Dong Nam/Kheua Kam Phii*	2285006	Respiratory	Reported by 1 PLM entry
bge099 (Euphorbiaceae) *Sauropus androgynus* (L.) Merr.	*Wan Ban* (*hak*)	2285007	Respiratory	Reported by 1 healer, 2 PLM entries
bge100 (Solanaceae) *Solanum lasiocarpum* Dunal	*Mak Euk/Mak Kheua Euk*	2285008	Respiratory	Reported by 3 healers
bge101 (Moringaceae) *Moringa oleifera* Lam.	*Ii Houm* (*hak*)	2282804	Respiratory	Reported by 3 PLM entries
bge102 (Amaranthaceae) *Amaranthus spinosus* L.	*Hom Nam/ Phak Hom* (*hak*)	2285000	Respiratory	Reported by 1 PLM entry
bge103 (Meliaceae) *Azadirachta indica* A. Juss.	*Khom Ka Dao*	2284999	Respiratory	Reported by 2 healers and 1 PLM entry
bge104 (Annonaceae) *Rollinia mucosa* (Jacq.) Baill.	*Khantaloht* (*peurk*)	2284998	Respiratory	Reported by 1 PLM entry
bge105 (Anacardiaceae) *Spondias* cf. *pinnata* (L. f.) Kurz	*Kok* (*mak/peurk*)	2285005	Respiratory	Reported by 1 PLM entry
bge106 (Moraceae) *Ficus* cf. *trichocarpa* Blume	*Deuah Kieng*	2284997	Respiratory	Reported by 1 PLM entry
bge107 (Polypodiaceae) *Drynaria quercifolia* (L.) J. Sm.	*Koot Hohk*	2285004	Respiratory	Reported by 1 healer
bge108 (Poaceae) *Saccharum officinarum* L.	*Oy Dam*	2285003	Respiratory	Reported by 3 healers and 23 PLM entries
bge110 (Rutaceae) *Feroniella lucida* Swingle	*Sung* (*mak/kok*)	2301516	Respiratory	Reported by 1 PLM entry
bge111 (Solanaceae) *Solanum cyanocarphium* Blume	*Mak Kheua Kheun* (*hak*)	2301564	Respiratory	Reported by 11 healers and 9 PLM entries
bge112 (Bignoniaceae) *Millingtonia hortensis* L. f.	*Kang Khong*	2301565	Respiratory	Reported by 11 healers and 7 PLM entries

(continued)

Table A.1 (continued)

Voucher herbarium specimen number (Family) Latin binomial	Common name	Field museum (F) Accession number	Use	Notes
bge113 (Annonaceae) *Marsypopetalum modestum* (Pierre) B. Xue & R.M.K. Saunders	*Tin Tang Tia*	2301566	Respiratory	Reported by 1 healer
bge114 (Rutaceae) *Micromelum minutum* Wight & Arn.	*Sa Mat Khao*	2301567	Respiratory	Reported by 1 healer
bge115 (Annonaceae) *Marsypopetalum modestum* (Pierre) B. Xue & R.M.K. Saunders	*Tin Tang Tia*	2301568	Respiratory	Reported by 1 healer
bge116 (Fabaceae) *Mucuna pruriens* (L.) DC.	*Tam Yay*	2301569	Respiratory	Reported by 5 healers and 5 PLM entries
bge117 (Bignoniaceae) *Fernandoa adenophylla* (Wall. ex G. Don) Steenis	*Kae Pa*	2301581	Respiratory	Reported by 2 healers
bge118 (Rutaceae) *Aegle marmelos* (L.) Corrêa	*Mak Toum*	2301570	Respiratory	Reported by 1 PLM entry
bge119 (Polygalaceae) *Securidaca inappendiculata* Hassk.	*Kheua Khao Mwak*	2301571	Respiratory	Reported by 1 healer
bge120 (Rubiaceae) *Benkara depauperata* (Drake) Ridsdale	*Kheua Khat Khao*	2301572	Respiratory	Reported by 2 PLM entries
bge121 (Moraceae) *Streblus asper* Lour.	*Som Phor*	2301573	Respiratory	Reported by 14 PLM entries
bge122 (Menispermaceae) *Tinospora crispa* (L.) Hook. f. & Thomson	*Kheua Khao Hor*	2301574	Respiratory	Reported by 2 healers and 17 PLM entries
bge123 (Annonaceae) *Uvaria cordata* Schumach. & Thonn.	*Tin Tang Tia*	2301575	Respiratory	Reported by 1 healer
bge124 (Annonaceae) *Anomianthus dulcis* (Dunal) James Sinclair	*Tin Tang Tia*	2301567	Respiratory	Reported by 1 healer
bge125 (Rutaceae) *Clausena harmandiana* (Pierre) Guillaumin	*Song Fa*	2301577	Respiratory, stomachic	Reported by 3 healers and 16 PLM entries

(continued)

Table A.1   (continued)

Voucher herbarium specimen number (Family) Latin binomial	Common name	Field museum (F) Accession number	Use	Notes
bge126 (Stemonaceae) *Stemona tuberosa* Lour.	*Kheua Hua Sam Sip/Kheuang Kheua*	2301578	Kidney stones	Reported by 1 healer
bge127 (Rubiaceae) *Mussaenda dinhensis* Pierre ex Pit.	*Tang Beu Kheua*	2301579	Nervous system disorders, back pain, tonic when mixed with ethanol	Reported by 1 healer
bge128 (Apocynaceae) *Cleghornia malaccensis* (Hook. f.) King & Gamble	*Kheua En Ohn*	2301543	Beri beri	Reported by 1 healer
bge129 (Cucurbitaceae) *Trichosanthes rubriflos* Thorel ex Cayla	*Kheua Mak Khi Ka*	2301544	Arthritis	Reported by 1 healer
bge130 (Rubiaceae) *Mussaenda dinhensis* Pierre ex Pit.	*Tang Beu*	2301545	Nervous system disorders, back pain, tonic when mixed with ethanol	Reported by 1 healer
bge131 (Loganiaceae) *Fagraea ceilanica* Thunb.	*Tang Tid Nok/Panya Meu Hoy*	2301546	Diabetes	Reported by 1 healer
bge132 (Annonaceae) *Uvaria rufa* Blume	*Mak Phii Phouan*	2301547	Stomachic and nervous system	Reported by 1 healer
bge133 (Moringaceae) *Moringa ovalifolia* Dinter & Berger	*Hak Ee Houm*	2301548	Diabetes	Reported by 1 healer
bge134 (Lamiaceae) *Gmelina philippensis* Cham.	*Kok Phoung Mool/Kok Phak Ka Deng*	2301549	Diabetes	Reported by 1 healer
bge135 (Loganiaceae) *Strychnos axillaris* Colebr.	*Sehn Pohng*	2301550	Tonic	Reported by 2 healers
bge137 (Annonaceae) *Uvaria rufa* Blume	*Tin Tang Tia*	2301551	Cough	Reported by 1 healer
bge138 (Polygalaceae) *Xanthophyllum lanceatum* J.J.Sm.	*Kok Seng*	2301515	Malaria	Reported by 2 healers

(continued)

Table A.1 (continued)

Voucher herbarium specimen number (Family) Latin binomial	Common name	Field museum (F) Accession number	Use	Notes
bge139 (Menispermaceae) *Pachygone* cf. *dasycarpa* Kurz	*Kheua Mak Hen*	2301552	Fever and arthritis	Reported by 2 healers
bge140 (Euphorbiaceae) *Actephila championiae* P.I. Forst.	*Kok Xii Nam*	2301514	Stomachic/ distended stomach	Reported by 2 healers
bge141 (Orchidaceae) *Cleisostoma* sp.	*Kap Kae Noy*	2301553	Lungs, asthma	Reported by 2 healers
bge142 (Rubiaceae) *Prismatomeris filamentosa* Craib	*Kok Ka Dan* (meh)	2301554	Liver disease	Reported by 2 healers
bge143 (Orchidaceae) *Coelogyne* sp.	*Ka Dam Phi*	2301555	Rash and black skin	Reported by 2 healers
bge148 (Pontederiaceae) *Monochoria hastata* (L.) Solms	*Phak Tohp*	2301556	Nervous system disorders	Reported by 2 healers
bge150 (Orchidaceae) *Rhynchostylis* sp.	*Kheua/Dok Kap Kae Nyai*	2301557	Freckle removal	Reported by 2 healers
bge151 (Loranthaceae) *Helixanthera parasitica* Lour.	*Phak Mai/Savan*	2301558	Skin disease	Reported by 2 healers
bge152 (Rubiaceae) *Cephalanthus tetrandra* (Roxb.) Ridsdale & Bakh. f.	*Kok Savan*	2301559	Skin disease	Reported by 2 healers
bge153 (Araceae) *Pistia stratiotes* L.	*Phak Johk Phou*	2301560	Skin disease	Reported by 2 healers
bge165 (Cyperaceae) *Actinoscirpus grossus* (L. f.) Goetgh. & D.A. Simpson	*Nya Pheu*	2301513	Back pain	Reported by 2 healers
bge173 (Lythraceae) *Rotala wallichii* (Hook. f.) Koehne	*Nya Ket Hoy/Phak Neh*	2301561	Skin disease	Reported by 2 healers
bge174 (Plantaginaceae) *Limnophila aromatica* (Lam.) Merr.	*Kha Nyeng, Phak/Ka Dien Nyai*	2301562	Nervous system disorders and distended stomach	Reported by 2 healers

(continued)

Table A.1 (continued)

Voucher herbarium specimen number (Family) Latin binomial	Common name	Field museum (F) Accession number	Use	Notes
bge176 (Poaceae) *Echinochloa* cf. *stagnina* (Retz.) P. Beauv.	*Nya Pong/Nya Sat Peuk*	2301512	Liver disease	Reported by 2 healers
bge181 (Euphorbiaceae) *Phyllanthus reticulatus* Poir.	*Am Ai*	2301563	Stomachic	Reported by 2 healers
bge186 (Balsaminaceae) *Hydrocera triflora* (L.) Wight & Arn.	*Ket Hoy Noy*	2301518	Skin disease	Reported by 2 healers
bge190 (Amaranthaceae) *Alternanthera sessilis* (L.) R. Br. ex DC.	*Nya Khohn Ta Xang/Phong Pheo/Nya Kha John*	2301519	Fever	Reported by 2 healers
bge192 (Poaceae) *Hygroryza* cf. *aristata* (Retz.) Nees ex Wright & Arn.	*Nya Pong Peng*	2301520	Rash	Reported by 2 healers
bge196 (Fabaceae) *Sesbania cannabina* (Retz.) Poir.	*Sa Noh Nam*	2301521	Tonic	Reported by 2 healers
bge198 (Menyanthaceae) *Nymphoides indica* (L.) Kuntze	*Phak Khanong Ma*	2301522	Stomachic	Reported by 2 healers
bge201 (Fabaceae) *Acacia pennata* (L.) Willd.	*Nam Han Khao/Kheua Han Khao*	2301511	Rash	Reported by 1 healer
bge202 (Simaroubaceae) *Harrisonia perforata* (Blanco) Merr.	*Kon Tha*	2301510	Back pain	Reported by 1 healer
bge203 (Euphorbiaceae) *Phyllanthus reticulatus* Poir.	*Am Ai*	2301509	Stomachic	Reported by 3 healers
bge204 (Lauraceae) *Litsea monopetala* (Roxb.) Pers.	*Kok Mii Hohm*	2301508	Fever, lice	Reported by 1 healer
bge205 (Rubiaceae) *Hedyotis macrosepala* (Pit.) P.H. Hô	*Ii Tou Pa*	2301507	Arthritis and tonic	Reported by 1 healer

(continued)

Table A.1 (continued)

Voucher herbarium specimen number (Family) Latin binomial	Common name	Field museum (F) Accession number	Use	Notes
bge206 (Combretaceae) *Combretum quadrangulare* Kurz	*Mak Kae*	2301506	Stomachic	Reported by 1 healer
bge207 (Rubiaceae) *Ixora balansae* Pit.	*Khem Nyai*	2301505	Fever	Reported by 1 healer
bge208 (Rutaceae) *Micromelum hirsutum* Merr.	*Sa Mat Dong/Sa Mat Nyai/Toht Deng*	2301523	Tonic	Reported by 1 healer
bge210 (Sterculiaceae) *Helicteres lanceolata* DC.	*Por Heen/Por Khii Kai Dam/Por Fan/Nya Khat* (meh)	2301504	Back pain	Reported by 1 healer
bge211 (Tiliaceae) *Colona auriculata* (Desf.) Craib	*Kheua Por Nam*	2301503	Stomachic	Reported by 1 healer
bge212 (Fabaceae) *Entada glandulosa* Pierre ex Gagnepain	*Kheua Ba Bohn*	2301524	Fever and jaundice	Reported by 2 healers
bge213 (Begoniaceae) *Begonia sinuata* Wall. ex Meisn.	*Phak Sohm Kung*	2301525	Stomachic	Reported by 1 healer
bge227 (Apocynaceae) *Holarrhena curtisii* King & Gamble	*Mouk Tia*	2301526	Stomachic	Reported by 1 healer
bge228 (Fabaceae) *Entada glandulosa* Pierre ex Gagnepain	*Kheua Ba Bohn*	2301527	Fever and jaundice	Reported by 2 healers
Bge232 (Rutaceae) *Clausena harmandiana* (Pierre) Guillaumin	*Song Fa*	2301528	Stomachic	Reported by 1 healer
bge237 (Euphorbiaceae) *Croton kongensis* Gagnep.	*Ma Ha Mek*	2301529	Fever	Reported by 1 healer
bge239 (Bignoniaceae) *Oroxylum indicum* (L.) Kurz	*Lin Mai*	2301530	Respiratory	Reported by 3 healers and 1 PLM entry

(continued)

Table A.1 (continued)

Voucher herbarium specimen number (Family) Latin binomial	Common name	Field museum (F) Accession number	Use	Notes
bge240 (Menispermaceae) *Tinospora crispa* (L.) Hook. f. & Thomson	*Kheua Khao Hor*	2301531	Respiratory	Reported by 2 healers and 17 PLM entries
bge241 (Annonaceae) *Uvaria* cf. *microcarpa* Champ. ex Benth.	*Phii Phouan*	2301532	Respiratory	Many healers identified a photo of *Marsypopetalum modestum* with this common name
bge243 (Verbenaceae) *Vitex trifolia* L.	*Phii Seua*	2301534	Respiratory	Reported by 7 PLM entries
bge244 (Menispermaceae) *Tinospora crispa* (L.) Hook. f. & Thomson	*Kheua Khao Hor*	2301535	Respiratory	Reported by 2 healers and 17 PLM entries
bge246 (Euphorbiaceae) *Sauropus androgynous* (L.) Merr.	*Phak Wan Ban*	2301536	Respiratory	Reported by 2 PLM entries
bge247 (Fabaceae) *Crotalaria pallida* Aiton	*Hing Hai*	2301537	Respiratory	Reported by 3 healers and 1 PLM entry
bge248 (Rutaceae) *Glycosmis pentaphylla* (Retz.) DC.	*Som Xeun*	2301538	Respiratory	Reported by 3 healers and 2 PLM entries
bge249 (Rutaceae) *Melicope* cf. *pteleifolia* (Champ. ex Benth.) T.G. Hartley	*Khom La Wan Joh*	2301539	Respiratory	Reported by 4 healers
bge250 (Lygodiaceae) *Lygodium microphyllum* (Cav.) R. Br.	*Koot Ngong*	2301540	Respiratory	Reported by 4 healers
bge251 (Simaroubaceae) *Irvingia malayana* Oliver ex Bennett	*Bohk*	2301541	Respiratory	Reported by 8 healers
bge252 (Rubiaceae) *Mitragyna* cf. *hirsuta* Havil.	*Thohm Phai*	2301542	Respiratory	Component of traditional remedy (TR1)
bge253 (Annonaceae) *Marsypopetalum modestum* (Pierre) B. Xue & R.M.K. Saunders	*Tin Tang Tia*	2301502	Respiratory	Recollected to make aqueous extract at UIC

(continued)

Table A.1 (continued)

Voucher herbarium specimen number (Family) Latin binomial	Common name	Field museum (F) Accession number	Use	Notes
bge254 (Bignoniaceae) *Millingtonia hortensis* L. f.	*Kang Khong*	2301501	Respiratory	Reported by 11 healers and 7 PLM entries
bge255 (Annonaceae) *Marsypopetalum modestum* (Pierre) B. Xue & R.M.K. Saunders	*Tin Tang Tia*	2304061	Respiratory	Flowering specimen recollected for identification
bge256 (Rutaceae) *Clausena harmandiana* (Pierre) Guillaumin	*Song Fa*	2304062	Respiratory	Reported by 3 healers and 16 PLM entries

The collection number represents the voucher herbarium specimen number. Subsets of this list were previously published by Elkington (2014a, b)

Appendix B
Glossary

English transliteration	Lao script	English translation
ajan	ອາຈານ	Respectful title for a teacher
aksep poht	ອັເສບປອດ	Lung infection
Am Ai	ອຳໃອ່	(Euphorbiaceae) *Phyllanthus reticulatus* Poir.
Ben Ah/En Ah	ເບັນອ້າ / ເອັນອ້າ	(Melastomataceae) *Melastoma malabathricum* L.
Bohk	ບົກ	(Simaroubaceae) *Irvingia malayana* Oliver ex Bennett
Bohk Khai	ບົກຄາຍ	(Euphorbiaceae) *Chaetocarpus castanocarpus* (Roxb.) Thwaites
Bou Ra	ບຸຣາ	Plant cited in the PLM—not identified
Deua Pong	ເດຶອປ່ອງ	(Moraceae) *Ficus erecta* Thunb.
Deuah Kieng	ເດຶອກ້ຽງ	(Moraceae) *Ficus* cf. *trichocarpa* Blume
hak	ຮາກ	Root
Hang Yen	ຮັງເຢັນ	(Fabaceae) *Millettia* sp.
Hing Hai	ທີງຫາຍ	(Fabaceae) *Crotalaria pallida* Aiton
Ho La Dan	ໂຫລະດານ	Plant cited in the PLM—not identified
Hom Nam/Phak Hom	ຫົມນຫາມ / ຜັກທົມ	(Amaranthaceae) *Amaranthus spinosus* L.
Hou Ha (kok)	ຮູຮາ (ກົກ)	(Acanthaceae) *Justicia adhatoda* L.
Ii Houm	ອີຮູມ	(Moringaceae) *Moringa oleifera* Lam.
Ii Tou Pa	ອີຕູ່ປ່າ	(Rubiaceae) *Hedyotis macrosepala* (Pit.) P.H. Hô
Jwang Hom	ຈວງທອມ	(Malvaceae) *Mansonia gagei* Drummond or (Rubiaceae) *Tarenna hoaensis* Pitard
Ka Dam Phi	ກາດາມຜີ	(Orchidaceae) *Coelogyne* sp.
Ka Thiem	ກະທຽມ	(Amaryllidaceae) *Allium sativum* L.

(continued)

© The Author(s) 2014

B.G. Elkington et al., *Ethnobotany of Tuberculosis in Laos*, SpringerBriefs in Plant Science, DOI 10.1007/978-3-319-10656-4

(continued)

English transliteration	Lao script	English translation
Kan Tong	ກ້ານຕົງ	(Rhamnaceae) *Colubrina javanica* Miq.
Kang Khong	ກາງຂອງ	(Bignoniaceae) *Millingtonia hortensis* L. f.
Kao Ho, Kheua	ເຂົາຮໍ, ເຄືອ	(Menispermaceae) *Tinospora crispa* (L.) Hook. f. & Thomson
Kao Jao	ເຂົ້າຈ້າວ	(Poaceae) *Oryza sativa* L. var. *dura*
Kap Kae Noy	ກັບແກ້ນ້ອຍ	(Orchidaceae) *Cleisostoma* sp.
Kasang (*mak*)	ກະສັງ (ໝາກ)	(Rutaceae) *Feroniella lucida* Swingle
Keuam (*kok*)	ເກີ່ມ (ກົກ)	(Burseraceae) *Canarium* cf. *hirsutum* Willd.
Kha	ຄ່າ	(Fabaceae) *Afzelia xylocarpa* (Kurz) Craib
Kha Leen	ຄ່າລິ້ນ	(Sapindaceae) *Dimocarpus longan* Lour.
Kha Nyeng, Phak	ກະແຍງ, ຜັກ	(Plantaginaceae) *Limnophila aromatica* (Lam.) Merr.
Khai Nao	ໄຂ່ເນົ່າ	(Rubiaceae) *Psychotria* sp.
Khai Nao (*noy*)	ໄຂ່ເນົ່ານ້ອຍ	(Rubiaceae) *Ixora* sp.
Khantaloht (*peurk*)	ຂັນທະລົດ (ເປືອກ)	(Annonaceae) *Rollinia mucosa* (Jacq.) Baill.
Kae Pa	ແຄປ່າ	(Bignoniaceae) *Fernandoa adenophylla* (Wall. ex G. Don) Steenis
Khem Nyai	ເຂັມໃຫຍ່	(Rubiaceae) *Ixora balansae* Pit.
Kheua Ba Bohn	ເຄືອບ້າບົນ	(Fabaceae) *Entada glandulosa* Pierre ex Gagnepain
Kheua En Ohn	ເຄືອເອັນອ່ອນ	(Apocynaceae) *Cleghornia malaccensis* (Hook. f.) King & Gamble
Kheua Hua Sam Sip/Kheuang Kheua	ເຄືອຫົວສາມສິບ / ເຂື່ອງເຄືອ	(Stemonaceae) *Stemona tuberosa* Lour.
Kheua Khao Mohk	ເຄືອຂາວມອກ	(Capparaceae) *Capparis* cf. *micrantha* A. Rich.
Kheua Khao Mwak	ເຄືອຂາວມວກ	(Polygalaceae) *Securidaca inappendiculata* Hassk.
Kheua Khat Khao	ເຄືອຂັດເຄົ້າ	(Rubiaceae) *Benkara depauperata* (Drake) Ridsdale
Kheua Mak Hen	ເຄືອໝາກແຫນ	(Menispermaceae) *Pachygone* cf. *dasycarpa* Kurz
Kheua Mak Khi Ka	ເຄືອໝາກຂີ້ກາ	(Cucurbitaceae) *Trichosanthes rubriflos* Thorel ex Cayla
Kheua Por Nam	ເຄືອປໍນ້ຳ	(Tiliaceae) *Colona auriculata* (Desf.) Craib
Kheua/Dok Kap Kae Nyai	ເຄືອ/ດອກກັບແກ້ໃຫຍ່	(Orchidaceae) *Rhynchostylis* sp.
Khii Fai Nok Khoum	ຂີ້ໄຟນົກຄຸ້ມ	(Asteraceae) *Elephantopus scaber* L.
Khii Lek	ຂີ້ເຫຼັກ	(Myrtaceae) *Decaspermum fruticosum* J.R. & G. Forster
Khing	ຂິງ	(Zingiberaceae) *Zingiber officinale* Roscoe
Khing Kheng	ຂິງແຄງ	(Zingiberaceae) *Curcuma parviflora* Wall.

(continued)

(continued)

English transliteration	Lao script	English translation
Khom Ka-Dao	ຂົມກະເດົາ	(Meliaceae) *Azadirachta indica* A. Juss.
Khom La Wan Joh	ຂົມລະຫວານຈໍ້	(Rutaceae) *Melicope* cf. *pteleifolia* (Champ. ex Benth.) T.G. Hartley
Khom Som	ຄອມສົ້ມ	(Tiliaceae) *Microcos paniculata* L.
Khor Phou (*mak*)	ກໍ່ພູ (ໝາກ)	(Meliaceae) *Sandoricum koetjape* (Burm. f.) Merr.
khwan	ຂວັນ	Spirit
kok	ກອກ	Tree trunk
Kok (*Peurk Mak Kok*)	ກອກ(ເປືອກໝາກກອກ)	(Anacardiaceae) *Spondias* cf. *pinnata* (L. f.) Kurz
Kok Ka Dan (meh)	ກົກກະດັນ (ແມ່)	(Rubiaceae) *Prismatomeris filamentosa* Craib
Kok Mii Hohm	ກົກໝີ້ຫອມ	(Lauraceae) *Litsea monopetala* (Roxb.) Pers.
Kok Phoung Moo/Kok Phak Ka Deng	ກົກຫຸງໝູ / ກົກຜັກກະແດງ	(Lamiaceae) *Gmelina philippensis* Cham.
Kok Savan	ກົກສະຫວານ	(Rubiaceae) *Cephalanthus tetrandra* (Roxb.) Ridsdale & Bakh. f.
Kok Seng	ກົກແສງ	(Polygalaceae) *Xanthophyllum lanceatum* J.J.Sm.
Kok Xii Nam	ກົກຊີ້ນຳ	(Euphorbiaceae) *Actephila championiae* P.I. Forst.
Kon Tha	ໂກນທາ	(Simaroubaceae) *Harrisonia perforata* (Blanco) Merr.
Koot Hohk	ກູດຮອກ	(Polypodiaceae) *Drynaria quercifolia* (L.) J. Sm.
Koot Ngong; Koot Khee Pa	ກູດງ້ອງ, ກູດຂີ້ປ່າ	(Lygodiaceae) *Lygodium microphyllum* (Cav.) R. Br. or *L. flexuosum* (L.) Sw.
Lan Xang	ລ້ານຊ້າງ	Million elephants
Lin Mai	ລີ້ນໄມ້	(Bignoniaceae) *Oroxylum indicum* (L.) Kurz
lom	ລົມ	Wind
Ma Ha Mek	ມະຫາເມກ	(Euphorbiaceae) *Croton kongensis* Gagnep.
mak	ໝາກ	Fruit
Mak Euk/Mak Kheua Euk	ໝາກເອິກ / ໝາກເຄືອເອິກ	(Solanaceae) *Solanum lasiocarpum* Dunal
Mak Kae	ໝາກແກ	(Combretaceae) *Combretum quadrangulare* Kurz
Mak Kheng Khom	ໝາກແກງຄົມ	(Molluginaceae) *Glinus oppositifolius* (L.) Aug. DC.
Mak Kheua Kheun	ໝາກເຂືອຂຶ້ນ	(Solanaceae) *Solanum cyanocarphium* Blume or *S. melongena* L.
Mak Phii Phouan	ໝາກພິພວນ	(Annonaceae) *Uvaria rufa* Blume
Mak Toum	ໝາກຕູມ	(Rutaceae) *Aegle marmelos* (L.) Corrêa

(continued)

(continued)

English transliteration	Lao script	English translation
Mao (mak)	ເໝົ້າ (ໝາກ)	(Euphorbiaceae) *Antidesma bunius* (L.) Spreng. or *A. diandrum* (Roxb.) Roth
Mouk Tia	ມູກເຕ້ັຍ	(Apocynaceae) *Holarrhena curtisii* King & Gamble
Nam Han Khao/Kheua Han Khao	ຫນາມຫັນຂາວ / ເຄືອຫັນຂາວ	(Fabaceae) *Acacia pennata* (L.) Willd.
Nam Lep Meo	ຫນາມເລັບແມວ	(Rhamnaceae) *Ziziphus oenopolia* (L.) Mill.
Nam Neh	ຫນາມແນ່	(Acanthaceae) *Thunbergia grandiflora* Roxb. or *Eranthemum pulchellum* Andrews
Niao Deng (mak)	ເຍົ້າແດງ (ໝາກ)	(Euphorbiaceae) *Jatropha gossypifolia* L.
Niao Khao (mak)	ເຍົ້າຂາວ (ໝາກ)	(Euphorbiaceae) *Jatropha curcas* L.
Nuat Meo Nam	ຫນວດແມວຫນາມ	(Lamiaceae) *Premna* sp.
Nya Boua	ຫຍ້າບົວ	(Cyperaceae) *Eleocharis spiralis* (Rottb.) Roem. & Schult.
Nya Ket Hoy/Phak Neh	ຫຍ້າເກີດຫອຍ / ຜັກແຫນ	(Lythraceae) *Rotala wallichii* (Hook. f.) Koehne
Nya Kha	ຫຍ້າຄາ	(Poaceae) *Imperata cylindrica* (L.) Beauvois
Nya Khohn Ta Xang/Phong Pheo/Nya Kha John	ຫຍ້າຂົນຕາຊ້າງ / ໂພງແພວ ຫຍ້າກະຈອນ	(Amaranthaceae) *Alternanthera sessilis* (L.) R. Br. ex DC.
Nya Lap Meun	ຫຍ້າລັບມືນ	(Fabaceae) *Cassia tora* L.
Nya Pheu	ຫຍ້າຜື	(Cyperaceae) *Actinoscirpus grossus* (L. f.) Goetgh. & D.A. Simpson
Nya Pong/Nya Sat Peuk	ຫຍ້າປ້ອງ / ຫຍ້າສັດເຜິກ	(Poaceae) *Echinochloa* cf. *stagnina* (Retz.) P. Beauv.
Oy Dam	ອ້ອຍດຳ	(Poaceae) *Saccharum officinarum* L.
Oy Sam Souan	ອ້ອຍສາມສວນ	(Euphorbiaceae) *Phyllanthus emblica* L.
Oy Xang	ອ້ອຍຊ້າງ	(Araliaceae) *Heteropanax fragrans* Seem.
Padong Khor/Pan Dong Nam/Kheua Kam Phii	ປະດົງຂໍ / ປັນດົງຫນາມ / ເຄືອກຳພີ້	(Fabaceae) *Dalbergia millettii* Benth.
peurk	ເປືອກ	Bark
Phak Johk Meh	ຜັກຈອກແມ່	(Salviniaceae) *Salvinia cucullata* Roxb.
Phak Johk Phou	ຜັກຈອກຜູ້	(Araceae) *Pistia stratiotes* L.
Phak Khanong Ma	ຜັກກະຫນ່ອງມ້າ	(Menyanthaceae) *Nymphoides indica* (L.) Kuntze
Phak Mai/Savan	ຜັກໄມ້ / ສະຫວັນ	(Loranthaceae) *Helixanthera parasitica* Lour.
Phak Nam	ຜັກຫນາມ	(Araceae) *Lasia spinosa* (L.) Thwaites
Phak Sohm Kung	ຜັກສົ້ມກຸ້ງ	(Begoniaceae) *Begonia sinuata* Wall. ex Meisn.
Phak Tohp	ຜັກຕົບ	(Pontederiaceae) *Monochoria hastata* (L.) Solms

(continued)

(continued)

English transliteration	Lao script	English translation
Phak Wan Ban	ຜັກຫວານບ້ານ	(Euphorbiaceae) *Sauropus androgynus* (L.) Merr.
Phet Pa (*mak*)	ເພັດປ່າ (ໝາກ)	(Apocynaceae) *Tabernaemontana bufalina* Lour.
Phii	ຜີ	Spirit
Phii Mob	ຜີມອບ	(Fabaceae) *Mimosa pudica* L. var. *hispida* Brenan
Phii Phouan	ຜີພວນ	(Annonaceae) *Uvaria* sp. or *Marsypopetalum modestum*
Phii Seua	ຜີເສື້ອ	(Verbenaceae) *Vitex trifolia* L.
Phohk	ພອກ	(Chrysobalanaceae) *Parinari anamensis* Hance
Pik (*mak*)	ພິກ (ໝາກ)	(Solanaceae) *Capsicum annuum* L. (fruit)
Pit Phii Deng	ປິດປີແດງ	(Plumbaginaceae) *Plumbago indica* L.
poht phiikan	ປອດພິການ	Deformed lungs
Por Heen/Por Khii Kai Dam/Por Fan/Nya Khat (meh)	ປໍຫິນ / ປໍຂີ້ໄກ່ດຳ / ປໍຟານ / ຫຍ້າຂັດ (ແມ່)	(Sterculiaceae) *Helicteres lanceolata* DC.
Poung Phing Deng	ພູງພິງແດງ	(Verbenaceae) *Clerodendrum palmatolobatum* Dop
Sa	ສາ	(Moraceae) *Broussonetia papyrifera* (L.) L'Hér. ex Vent.
Sa Mat	ສະມັດ	(Rutaceae) *Micromelum falcatum* (Lour.) Tanaka
Sa Mat Dong/Sa Mat Nyai/Toht Deng	ຕົດແດງ	(Rutaceae) *Micromelum hirsutum* Merr.
Sa Mat Khao	ສະມັດຂາວ	(Rutaceae) *Micromelum minutum* Wight & Arn.
Sa Noh Nam	ສະໂນບນ້ຳ	(Fabaceae) *Sesbania cannabina* (Retz.) Poir.
sainyasat	ໄຊຍະລາດ, ໄສຍະສາດ	Magic. A more thorough explanation is given by McDaniel (2011)
Sam Sip (*hua*)	ຫົວສາມສິບ	(Stemonaceae) *Stemona cochinchinensis* Gagnep. or *S. tuberosa* Lour.
Sehn Pohng	ຮສບປ້ອງ	(Loganiaceae) *Strychnos axillaris* Colebr.
Sii Khai	ສີໄຄ	(Poaceae) *Cymbopogon nardus* Rendle
Sii Khai Tohn	ສີໄຄຕົ້ນ	(Lauraceae) *Litsea cubeba* (Lour.) Pers. or *Cinnamomum iners* Reinw. ex Blume
Som Kop	ສົ້ມກົບ	(Rubiaceae) *Hymenodictyon orixense* (Roxb.) Mabb.
Som Lom	ສົ້ມລົ້ມ	(Apocynaceae) *Aganonerion polymorphum* Pierre in Spire & A. Spire or *Ecdysanthera rosea* Hook. & Arn.
Som Phor	ສົ້ມພໍ່	(Moraceae) *Streblus asper* Lour.
Som Phory	ສົ້ມປ່ອຍ	(Fabaceae) *Acacia leucophloea* Willd.

(continued)

(continued)

English transliteration	Lao script	English translation
Som Xeun	ສົ້ມຂື່ນ	(Rutaceae) *Glycosmis pentaphylla* (Retz.) DC. or *G. cochinchinensis* (Lour.) Pierre
Song Fa	ສ່ອງຟ້າ	(Rutaceae) *Clausena harmandiana* (Pierre) Guillaumin
Tah Kai	ຕາໄກ່	(Celastraceae) *Salacia chinensis* L.
Tam Yay	ຕຳແຍ	(Fabaceae) *Mucuna pruriens* (L.) DC.
Tang Beu	ຕັ່ງເບື	(Rubiaceae) *Mussaenda dinhensis* Pierre ex Pit.
Tang Tid Nok/Panya Meu Hoy	ຕັ່ງຕິດນົກ / ພະຍາມືຫອຍ	(Loganiaceae) *Fagraea ceilanica* Thunb.
Tao Hang	ຕາວຫ້າງ	(Arecaceae) *Arenga caudata* (Lour.) H.E. Moore
Thohm Phai	ທົມພາຍ	(Rubiaceae) *Mitragyna rotundifolia* (Roxb.) Kuntze or *M. hirsuta* Havil.
Tin Cham Kohn	ຕີນຈຳຂົນ	(Myrsinaceae) *Ardisia* sp.
Tin Nohk	ຕີນນົກ	(Araliaceae) *Schefflera* sp.
Tin Tang Tia	ຕີນຕັ່ງເຕັ້ຍ	(Annonaceae) *Marsypopetalum modestum* or *Uvaria* sp.
Toum Kah Khao	ຕູມກາຂາວ	(Loganiaceae) *Strychnos nux-blanda* A.W. Hill
wat	ວັດ	Buddhist temple or pagoda
Xai Xou Tohn (*hak*)	ຊາຍຊູ້ຕົ້ນ (ຮາກ)	(Capparaceae) *Capparis micrantha* A. Rich.
Ya Hua	ຍາຫົວ	(Smilacaceae) *Smilax glabra* Roxb.

The scientific names for these plants are the result of taxonomic identification. In some cases, other research has identified different scientific names for the same common names, such as in Inthakoun and Delang (2011)

Appendix C
Select Plant Images (in Alphabetical Order by Family—Genus—Species Names)

Fig. C.1 Acanthaceae—*Justicia* cf. *adhatoda* L.

Kok Hou Ha
ກົກຮູຮາ

B.G. Elkington et al., *Ethnobotany of Tuberculosis in Laos*, SpringerBriefs
in Plant Science, DOI 10.1007/978-3-319-10656-4

71

Fig. C.2 Amaranthaceae—*Alternanthera sessilis* (L.) R. Br. ex DC.

Nya Khohn Ta Xang / Phong Pheo / Nya Kha John
ຫຍ້າຂົ້ນຕາຊ້າງ / ໂພງແພວ / ຫຍ້າກະຈອນ

Fig. C.3 Amaranthaceae—*Amaranthus spinosus* L.

Hom Nam / Phak Hom
ຫົມຫນາມ / ຜັກຫົມ

Fig. C.4   Anacardiaceae—*Spondias* cf. *pinnata* (L. f.) Kurz

Kok (*Peurk Mak Kok*)
ກອກ(ເປືອກໝາກກອກ)

Fig. C.5   Annonaceae—*Uvaria rufa* Blume

Mak Phii Phouan / *Tin Tang Tia*
ໝາກພີພວນ / ຕີນຕັ່ງເຕັ້ຍ

Fig. C.6 Apocynaceae—*Cleghornia malaccensis* (Hook. f.) King & Gamble

Kheua En Ohn
ເຄືອເອັນອ່ນ

Fig. C.7 Apocynaceae—*Holarrhena curtisii* King & Gamble

Mouk Tia
ມູກເຕ້ຍ

Fig. C.8 Araceae—*Lasia spinosa* (L.) Thwaites

Phak Nam
ຜັກຫນາມ

Fig. C.9 Araceae—*Pistia stratiotes* L.

Phak Johk Phou
ຜັກຈອກຜູ້

Fig. C.10 Arecaceae—*Arenga caudata* (Lour.) H.E. Moore

Tao Hang
ຕາວຫາງ

Fig. C.11 Asteraceae—*Elephantopus scaber* L.

Khii Fai Nok Khoum
ຂີ້ໄຟນົກຄຸ້ມ

Fig. C.12 Begoniaceae—*Begonia sinuata* Wall. ex Meisn.

Phak Sohm Kung
ຜັກສົ້ມກຸ້ງ

Fig. C.13 Bignoniaceae—*Fernandoa adenophylla* (Wall. ex G.Don) Steenis

Kae Pa
ແຄປ່າ

Fig. C.14 Bignoniaceae—*Millingtonia hortensis* L. f.

Kang Khong
ກາງຂອງ

Fig. C.15 Bignoniaceae—*Oroxylum indicum* (L.) Kurz

Lin Mai
ລິ້ນໄມ້

Fig. C.16 Burseraceae—*Canarium* cf. *hirsutum* Willd.

Kok Kheuam
ກົກເຄືມ

Fig. C.17 Chrysobalanaceae—*Parinari anamensis* Hance

Phohk
ພອກ

Fig. C.18 Combretaceae—*Combretum quadrangulare* Kurz

Mak Kae
ໝາກແກ

Fig. C.19 Cucurbitaceae—*Trichosanthes rubriflos* Thorel ex Cayla

Kheua Mak Khi Ka
ເຄືອໝາກຂີ້ກາ

Fig. C.20 Cyperaceae—*Actinoscirpus grossus* (L. f.) Goetgh. & D.A. Simpson

Nya Pheu
ຫຍ້າຜີ

Fig. C.21 Euphorbiaceae—*Actephila championiae* P.I. Forst

Kok Xii Nam

ກົກຊີ້ຫນາມ

Fig. C.22 Euphorbiaceae—*Antidesma bunius* (L.) Spreng.

Mao (mak)

ເໝົ້າ (ໝາກ)

Fig. C.23 Euphorbiaceae—*Chaetocarpus castanocarpus* (Roxb.) Thwaites

Bohk Khai

ບົກຄາຍ

Fig. C.24 Euphorbiaceae—*Croton kongensis* Gagnep.

Ma Ha Mek
ມະຫາເມກ

Fig. C.25 Euphorbiaceae—*Jatropha gossypifolia* L.

Niao Deng (mak)
ເຍົາແດງ (ໝາກ)

Fig. C.26 Euphorbiaceae—*Phyllanthus emblica* L.

Oy Sam Souan
ອອ້ຍສາມສວນ

Fig. C.27 Euphorbiaceae—*Phyllanthus reticulatus* Poir.

Am Ai
ອຳໃອ່

Fig. C.28 Euphorbiaceae—*Sauropus androgynus* (L.) Merr.

Phak Wan Ban
ຜັກຫວານບ້ານ

Fig. C.29 Fabaceae - Mimosoideae—*Acacia pennata* (L.) Willd.

Nam Han Khao / Kheua Han Khao
ຫນາມຫັນຂາວ / ເຄືອຫັນຂາວ

Fig. C.30 Fabaceae - Mimosoideae—*Entada glandulosa* Pierre ex Gagnepain

Kheua Ba Bohn
ເຄືອບ້າບົນ

Fig. C.31 Fabaceae - Papilionoideae—*Mucuna pruriens* (L.) DC.

Tam Yay
ຕຳແຍ

Fig. C.32 Lamiaceae—*Gmelina philippensis* Cham.

Kok Phoung Moo / Kok Phak Ka Deng
ກົກພຸງໝູ / ກົກຜັກກະແດງ

Fig. C.33 Lamiaceae—*Premna* sp.

Nuat Meo Nam
ໜວດແມວໝາມ

Fig. C.34 Loganiaceae—*Fagraea ceilanica* Thunb.

Tang Tid Nok / Panya Meu Hoy
ຕັ້ງຕິດນົກ / ພະຍາມືຫອຍ

Fig. C.35 Loganiaceae—*Strychnos axillaris* Colebr.

Sehn Pohng
ະສນປ້ອງ

Fig. C.36 Loranthaceae—*Helixanthera parasitica* Lour.

Phak Mai / Savan
ຜັກໄມ້ / ສະຫວັນ

Fig. C.37 Lythraceae—*Rotala wallichii* (Hook. f.) Koehne

Nya Ket Hoy / Phak Neh
ຫຍ້າເກັດຫອຍ / ຜັກແຫນ

Fig. C.38 Melastomataceae—*Melastoma malabathricum* L.

Ben Ah / En Ah
ເບັນອ້າ / ເອັນອ້າ

Fig. C.39 Menispermaceae—*Tiliacora* cf. *triandra* Diels

Kheua Mak Hen
ເຄືອໝາກເຫັນ

Fig. C.40 Menispermaceae—*Tinospora crispa* (L.) Hook. f. & Thomson

Kheua Khao Hor
ເຄືອເຂົາຫໍ

Fig. C.41 Menyanthaceae—*Nymphoides indica* (L.) Kuntze

Phak Khanong Ma
ຜັກຂະໜ່ອງມ້າ

Fig. C.42 Moraceae—*Ficus* cf. *trichocarpa* Blume

Deuah Kieng
ເດືອກ຺ຽງ

Fig. C.43 Moraceae—*Ficus* cf. *erecta* Thunb.

Deua Pong
ເດືອປ່ອງ

Fig. C.44 Moringaceae—*Moringa oleifera* Lam.

Ii Houm
ອິຮູ່ມ

Fig. C.45 Orchidaceae—*Cleisostoma* sp.

Kap Kae Noy
ກັບແກ້ນ້ອຍ

Fig. C.46 Orchidaceae—*Coelogyne* sp.

Ka Dam Phi
ກາດຳຜິ

Fig. C.47 Orchidaceae—*Rhynchostylis* sp.

Kheua / Dok Kap Kae Nyai
ເຄືອ / ດອກກັບແກ້ໃຫຍ່

Fig. C.48 Poaceae—*Echinochloa* cf. *stagnina* (Retz.) P. Beauv.

Nya Pong / Nya Sat Peuk
ຫຍ້າປ້ອງ / ຫຍ້າສັດເຜິກ

Fig. C.49 Polygalaceae—*Xanthophyllum lanceatum* J.J.Sm.

Kok Seng
ກົກແສງ

Fig. C.50 Polypodiaceae—*Drynaria quercifolia* (L.) J. Sm.

Koot Hohk
ກູດຮອກ

Fig. C.51 Pontederiaceae—*Monochoria hastata* (L.) Solms

Phak Tohp
ຜັກຕົບ

Fig. C.52 Rhamnaceae—*Colubrina javanica* Miq.

Kan Tong
ກ້ານຕົ໌ງ

Fig. C.53 Rhamnaceae—*Ziziphus oenopolia* (L.) Mill.

Nam Lep Meo
ໜາມເລັບແມວ

Fig. C.54 Rubiaceae—*Benkara depauperata* (Drake) Ridsdale

Kheua Khat Khao
ເຖືອຂັດເຖົ້າ

Fig. C.55 Rubiaceae—*Cephalanthus tetrandra* (Roxb.) Ridsdale & Bakh. f.

Kok Savan
ກົກສະອານ

Fig. C.56   Rubiaceae—*Hedyotis macrosepala* (Pit.) P.H. Hô

Ii Tou Pa
ອີ່ຕູ່ປ່າ

Fig. C.57   Rubiaceae—*Ixora balansae* Pit.

Khem Nyai
ເຂັມໃຫຍ່

Fig. C.58   Rubiaceae—*Mussaenda dinhensis* Pierre ex Pit.

Tang Beu
ຕັງເບື

Fig. C.59 Rubiaceae—*Prismatomeris filamentosa* Craib

Kok Ka Dan (meh)
ກົກກະດັນ (ແມ່)

Fig. C.60 Rutaceae—*Aegle marmelos* (L.) Corrêa

Mak Toum
ໝາກຕູມ

Fig. C.61 Rutaceae—*Clausena harmandiana* (Pierre) Guillaumin

Song Fa

ສ່ອງຟ້າ

Fig. C.62 Rutaceae—*Feroniella lucida* Swingle

Kasang (*mak*)

ກະສັງ (ໝາກ)

Fig. C.63 Rutaceae—*Glycosmis pentaphylla* (Retz.) DC.

Som Xeun

ສົ້ມຊື່ນ

Fig. C.64 Rutaceae—*Micromelum* cf. *falcatum* (Lour.) Tanaka

Sa Mat
ສະມັດ

Fig. C.65 Rutaceae—*Micromelum minutum* Wight & Arn.

Sa Mat Khao
ສະມັດຂາວ

Fig. C.66 Salviniaceae—*Salvinia cucullata* Roxb.

Phak Johk Meh

ຜັກຈອກແມ່

Fig. C.67 Simaroubaceae—*Harrisonia perforata* (Blanco) Merr.

Kon Tha

ໂກນທາໆ

Fig. C.68 Simaroubaceae—*Irvingia malayana* Oliver ex Bennett

Bohk

ບົກ

Fig. C.69   Solanaceae—*Solanum lasiocarpum* Dunal

Mak Euk / Mak Kheua Euk
ໝາກເອິກ / ໝາກເຄືອເອິກ

Fig. C.70   Stemonaceae—*Stemona cochinchinensis* Gagnep.

Hua Sam Sip
ຫົວສາມສິບ

Fig. C.71 Stemonaceae—*Stemona tuberosa* Lour.

Kheua Hua Sam Sip / Kheuang Kheua
ເຄືອຫົວສາມສິບ / ເຂືອງເຄືອ

Fig. C.72 Sterculiaceae—*Helicteres lanceolata* DC.

Por Heen / Por Khii Kai Dam / Por Fan / Nya Khat (*meh*)
ປໍຫິນ / ປໍຂີ້ໄກ່ດຳ / ປໍຟານ / ຫຍ້າຂັດ (ແມ່)

Fig. C.73 Tiliaceae—*Colona auriculata* (Desf.) Craib

Kheua Por Nam
ເຄືອປໍນ້ຳ

Fig. C.74 Tiliaceae—*Microcos paniculata* L.

Khom Som
ຄອມສົ້ມ

Fig. C.75 Verbenaceae—*Clerodendrum palmatolobatum* Dop

Poung Phing Deng
ພວງພິງແດງ

Fig. C.76 Verbenaceae—*Vitex trifolia* L.

Phii Seua
ຜີເສື້ອ

References

Abhakorn R (1997) Towards a collective memory of mainland Southeast Asia: field preservation of traditional manuscripts in Thailand, Laos and Myanmar. In: Manning RW, Kremp V (eds) A reader in perservation and conservation. International Federation of Library Associations, Munchen, pp 86–91

ADB (2007) Fact sheet on Lao PDR, 2007. Asian Development Bank, Vientiane

ADB (2009) The Lao People's Democratic Republic. Asian Development Bank, Mandaluyong

ADB (2012) Overview: greater Mekong subregion, 2012. Asian Development Bank, Mandaluyong. http://www.adb.org/countries/gms/overview. Accessed 3 Mar 2012

Adkins JE, Boyer EW, McCurdy CR (2011) *Mitragyna speciosa*, a psychoactive tree from Southeast Asia with opioid activity. Curr Top Med Chem 11(9):1165–1175

Alexiades MN (1996) Selected guidelines for ethnobotanical research: a field manual. In: Sheldon JW (ed) Advances in economic botany. New York Botanical Garden, New York

Baird IG (2011) The Don Sahong Dam. Critical Asian Studies 43(2):211–235

Balick MJ (1990) Ethnobotany and the identification of therapeutic agents from the rainforest. In: Chadwick D, Marsh J (eds) Bioactive compounds from plants, vol 154. Wiley, Chichester

Barger G, Dyer E, Sargent LJ (1939) The alkaloids of *Mitragyna rotundifolia*. J Org Chem 04(4):418–427

Campbell IC (2009) Introduction. In: Campbell IC (ed) The Mekong: biophysical environment of an international river Basin. Elsevier, New York

Campinha-Bacote J (2007) Becoming culturally competent in ethnic psychopharmacology. J Psychosoc Nurs 45(9):26–33

Cantrell CL, Lu T, Fronczek FR, Fischer NH, Adams LB, Franzblau SG (1996) Antimycobacterial cycloartanes from *Borrichia frutescens*. J Nat Prod 59(12):1131–1136

Case R (2006) Integrative pharmacognostic evaluation of anti-TB ethnobotanicals from manus. PhD Dissertation. University of Illinois at Chicago, Chicago

Chazee L (1999) The peoples of Laos: rural and ethnic diversities. White Lotus, Bangkok

Cho SH, Warit S, Wan B, Hwang CH, Pauli GF, Franzblau SG (2007) Low-oxygen-recovery assay for high-throughput screening of compounds against nonreplicating *Mycobacterium tuberculosis*. Antimicrob Agents Chemother 51(4):1380–1385

CIA (February 5, 2013) World Factbook: Laos. http://www.cia.gov/library/publications/the-world-factbook/geos/la.html. Accessed 18 Feb 2013

CDC (2012) Tuberculosis (TB): drug-resistant TB. Centers for Disease Control and Prevention, Atlanta. http://www.cdc.gov/tb/topic/drtb/default.htm Accessed 16 Oct 2014

© The Author(s) 2014

B.G. Elkington et al., *Ethnobotany of Tuberculosis in Laos*, SpringerBriefs in Plant Science, DOI 10.1007/978-3-319-10656-4

Collins L, Franzblau SG (1997) Microplate alamar blue assay versus BACTEC 460 system for high-throughput screening of compounds against *Mycobacterium tuberculosis* and *Mycobacterium avium*. Antimicrob Agents Chemother 41(5):1004–1009

Cragg GM, Boyd MR, Cardellina JH 2nd, Newman DJ, Snader KM, McCloud TG (1994) Ethnobotany and drug discovery: the experience of the US National Cancer Institute. Ciba Found Symp 185:178–190, discussion 190-176

Cummings J, Burke A (2005) Laos. Lonely Planet, Oakland

de Zoysa N (2000) Arecaceae. In: Dassanayake MD, Clayton WD (eds) A revised handbook to the Flora of Ceylon, vol 14. A.A. Balkema, Brookfield

Drugs.com (2012) Dipyrithione. http://www.drugs.com/international/dipyrithione.html. Accessed 18 Feb 2013

Elkington BG (2013) Traditional herbal treatments for tuberculosis in Laos: ethnobotany and pharmacognosy studies. University of Illinois at Chicago, Chicago

Elkington B, Southavong B, Sydara K, Souliya O, Vanthanouvong M, Nettavong K, Thammachack B, Pak DH, Riley MC, Franzblau SG, Soejarto DD (2009) Biological evaluation of plants of Laos used in the treatment of tuberculosis in Lao traditional medicine. Pharm Biol 47(1): 26–33

Elkington B, Southavong B, Sydara K, Bouamanivong S, Souliya O, Xayveu M, Vanthanouvong M, Riley M, Panyachit B, Thammachak B, Soejarto DD (2010) The search for anti-malarial plants in Lao palm-leaf manuscripts. In: Adams K, Hudak TJ, Tempe AZ (eds) Multidisciplinary perspectives on Lao studies. Southeast Asia Council Center for Asian Research Arizona State University, Tempe, pp 277–288

Elkington B, Sydara K, Hartmann JF, Southavong B, Soejarto DD (2012) Folk epidemiology recorded in palm leaf manuscripts of Laos. J Lao Studies 3:1–14

Elkington BG, Phiapalath P, Sydara K, Somsamouth V, Goodsmith NI, Soejarto DD (2014a) Assessment on the importance of medicinal plants among communities around Khiat Ngong of southern Laos. J Environ Biol 35:607–615

Elkington BG, Sydara K, Newsome A, Hwang CH, Lankin DC, Simmler C, Napolitano JG, Ree R, Graham JG, Gyllenhaal C, Bouamanivong S, Souliya O, Pauli GF, Franzblau SG, Soejarto DD (2014b) New finding of an anti-TB compound in the genus *Marsypopetalum* (Annonaceae) from a traditional herbal remedy of Laos. J Ethnopharmacol 151(2):903–911

Enfield NJ (2003) Linguistic epidemiology: semantics and grammar of language contact in mainland Southeast Asia. Routledge Curzon, London

Enfield NJ (2006) Laos—language situation. In: Brown K (ed) Encyclopedia of language and linguistics, vol 6. Elsevier, Amsterdam, pp 698–700

Enfield NJ, Evans G (2000) Transcription as standardisation: the problem of Tai languages. In: Burusphat S (ed) Paper presented at the international conference on Tai studies, Bangkok, Thailand, 29–31 July 1998

Evans G (1995) What is Lao culture and society? In: Evans G (ed) Laos: culture and society. Silkworm Books, Chiang Mai

Evans G (2002) A short history of Laos: the land in between. In: Osborne M (ed) Short history of Asia series. Allen & Unwin, Crows Nest

Fadiman A (1998) The sprit catches you and you fall down: a Hmong child, her American doctors, and the collision of two cultures. Farrar, Straus and Giroux, New York

Falzari K, Zhu Z, Pan D, Liu H, Hongmanee P, Franzblau SG (2005) *In vitro* and *in vivo* activities of macrolide derivatives against *Mycobacterium tuberculosis*. Antimicrob Agents Chemother 49(4):1447–1454

Flora of China (1994a) Broussonetia papyrifera. Flora of China. Vol. 5. Science Press & Missouri Botanical Garden, Beijing. http://www.efloras.org/florataxon.aspx?flora_id=2&taxon_id=200006341. Accessed 13 Jan 2013

Flora of China (1994b) Tinospora crispa. Flora of China. Vol. 7. Science Press & Missouri Botanical Garden, Beijing. http://www.efloras.org/florataxon.aspx?flora_id=2&taxon_id=242352330. Accessed 18 Feb 2013

Franzblau SG, Witzig RS, McLaughlin JC, Torres P, Madico G, Hernandez A, Degnan MT, Cook MB, Quenzer VK, Ferguson RM, Gilman RH (1998) Rapid, low-technology MIC determination with clinical *Mycobacterium tuberculosis* isolates by using the microplate Alamar Blue assay. J Clin Microbiol 36(2):362–366

Gorman C, Charoenwongsa P (1976) Ban Chiang—a mosaic of impressions from the first two years. Expedition 18(4):14–26

Grabowsky V (1995) The Isan up to its integration into the Siamese state. In: Grabowsky V (ed) Regions and national integration in Thailand 1892–1992. Harrassowitz, Wiesbaden

Grange JM, Snell NJ (1996) Activity of bromhexine and ambroxol, semi-synthetic derivatives of vasicine from the Indian shrub *Adhatoda vasica*, against *Mycobacterium tuberculosis in vitro*. J Ethnopharmacol 50(1):49–53

Gupta PR (2010) Ambroxol—resurgence of an old molecule as an anti-inflammatory agent in chronic obstructive airway diseases. Lung India 27(2):46–48

Gutierrez MC, Brisse S, Brosch R, Fabre M, Omais B, Marmiesse M, Supply P, Vincent V (2005) Ancient origin and gene mosaicism of the progenitor of *Mycobacterium tuberculosis*. PLoS Pathog 1(1):e5

Gyllenhaal C, Kadushin MR, Soejarto DD, Southavong B, Sydara K, Bouamanivong S, Xayveu M, Xuan L, Hiep N, Hung N, Loc P, Dac L, Binh L, Hai N, Bich T, Cuong N, Zhang H, Franzblau SG, Xie H, Riley MC, Elkington B, Nguyen HT, Waller DP, Tamez P, Tan G, Pezzuto J (2012) Ethnobotanical approach versus random approach in the search for new bioactive compounds: support of a hypothesis. Pharm Biol 50(1):30–41

Han C, Fu J, Liu Z, Huang H, Luo L, Yin Z (2010) Dipyrithione inhibits IFN-γ-induced JAK/STAT1 signaling pathway activation and IP-10/CXCL10 expression in RAW264.7 cells. Inflamm Res 59(10):809–816

Hartmann JF (1986) The spread of South Indic scripts in Southeast Asia. Crossroads 3(1):6–20

Hartmann JF (2002) Spoken Lao—a regional approach. 2003 SEAsite Laos. Northern Illinois University, DeKalb. http://www.seasite.niu.edu/lao/lao3.htm. Accessed 6 Dec 6 2011

Hazra KL (1994) Pāli language and literature: a systematic survey and historical study, Emerging perceptions in Buddhist studies; no. 4–5. D.K. Printworld, New Delhi

Higham C (2002) Early cultures of mainland Southeast Asia. River Books, Bangkok

Ho PH (1993) Cay Co Viet Nam: An illustrated Flora of Viet Nam, vol 3. Mekong Printing, Montreal

Huang H, Pan Y, Ye Y, Gao M, Yin Z, Luo L (2011) Dipyrithione attenuates oleic acid-induced acute lung injury. Pulm Pharmacol Ther 24(1):74–80

Hundius H (2005) Lao manuscripts and traditional literature: the struggle for their survival. The literary heritage of Laos: preservation, dissemination and research perspectives. National Library of Laos, Vientiane

Hundius H, Wharton D (2011) The digital library of Lao Manuscripts. J Lao Studies 2(2):67–74

Inthakoun L, Delang CO (2011) Lao Flora: a checklist of plants found in Lao PDR with scientific and vernacular names. Lulu Enterprises, Morrisville

IPNI (2008) Plant name details: Menispermaceae *Tinospora crispa* (L.) Hook.f. & Thomson. The international plant names index. The Royal Botanic Gardens, Kew, The Harvard University Herbaria, and the Australian National Herbarium. http://www.ipni.org/ipni/idPlantNameSearch.do?id=581600-1. Accessed 18 Feb 2013

ISE (2006) International society of ethnobiology code of ethics (with 2008 additions). http://ethnobiology.net/docs/ISECOE_Eng_rev_24Nov08.pdf. Accessed 19 Feb 2013

IUCN (2012) IUCN red list of threatened species. Version 2012.2. http://www.iucnredlist.org/. Accessed 19 Feb 2013

Kang WY, Zhang BR, Xu QT, Li L, Hao XJ (2006) Study on the chemical constituents of *Mitragyna rotundifolia*. Zhong Yao Cai 29(6):557–560

Kent C (1992) Perspectives. Tuberculosis: the "white plague" rises. Faulkner Grays Med Health 46(35):suppl 4 p

Kleinman A, Eisenberg L, Good B (1978) Culture, illness, and care: clinical lessons from anthropologic and cross-cultural research. Ann Intern Med 88(2):251–258

Koehler CSW (2002) Consumption, the great killer. Modern Drug Discovery 5(2):47–49

Kwong SL, Stewart SL, Aoki CA, Chen MS Jr (2010) Disparities in hepatocellular carcinoma survival among Californians of Asian ancestry, 1988 to 2007. Cancer Epidemiol Biomarkers Prev 19(11):2747–2757

Lazarus K, Dubeau P, Bambaradeniya C, Friend R, Sylavong L (2006) An uncertain future: biodiversity and livelihoods along the Mekong river in Northern Lao PDR. IUCN, Bangkok

Lewis PM (2009) Ethnologue: languages of the world, 16th edn. SIL International, Dallas

Lewis WH, Elvin-Lewis MP (1995) Medicinal plants as sources of new therapeutics. Ann Mo Bot Gard 82(1):16–24

Libman AS, Southavong B, Sydara K, Bouamanivong S, Gyllenhaal C, Riley MC, Soejarto DD (2009) The influence of cultural tradition and geographic location on the level of medicinal plant knowledge held by various cultural groups in Laos. In: Compton CJ, Hartmann JF, Sysamouth V (eds) Contemporary Lao studies: research on development, language and culture, and traditional medicine. Southeast Asia Publications, DeKalb, p 339

Lorrillard M (2006) Lao history revisited: paradoxes and problems in current research. South East Asia Res 14(3):387–401

LSB (2005) Population census Chapter 2: population census—population distribution and migration. Lao Statistics Bureau. http://nsc.gov.la/index.php?option=com_content&view=article&id=18&Itemid=19&lang=en&limitstart=1. Accessed 19 Feb 2013

Lu Z, Dai S, Chen R, Yu D (2010) 2-Pyridinethiolate derivatives from branches and leaves of *Polyalthia nemoralis* and their cytotoxic activities. Zhongguo Zhong Yao Za Zhi 35(1):53–57

Łuczaj ŁJ (2010) Plant identification credibility in ethnobotany: a closer look at Polish ethnographic studies. J Ethnobiol Ethnomed 6(1):36

Matisoff J (1991) Areal and universal dimensions of grammatization in Lahu. In: Traugott EC, Heine B (eds) Approaches to grammaticalization, vol 2. Benjamins, Amsterdam, pp 383–453

McCarthy JFRGS (1900) Surveying and exploring in Siam. J. Murray, London

McCloskey LA, Southwick K (1996) Psychosocial problems in refugee children exposed to war. Pediatrics 97(3):394–397

McDaniel JT (2008) Gathering leaves & lifting words: histories of Buddhist monastic education in Laos and Thailand. University of Washington, Seattle

McDaniel JT (2011) The Lovelorn Ghost and the Magical Monk: practicing Buddhism in modern Thailand. Columbia University, New York

McDonald NA (1999) A missionary in Siam (1860–1870). White Lotus, Bangkok

Merck (2009) Tuberculosis (TB). In: Nardell EA (ed) The Merck manual. Merck, Whitehouse Station

MOIC (2012) Diagnostic trade integration study. Department of Planning and Cooperation, Ministry of Industry and Commerce, Lao PDR, Vientiane

MRC (2012) The Mekong Basin: physiography. Mekong River Commission, Vientiane. http://www.mrcmekong.org/the-mekong-basin/physiography. Accessed 18 Feb 2013

Murray JF (2004) A century of tuberculosis. Am J Respir Crit Care Med 169(11):1181–1186

NAPRALERT (2010) *Tinospora crispa*: common names and latin synonyms. Program for Collaborative Research in the Pharmaceutical Sciences. University of Illinois at Chicago, Chicago

NBSAP (2004) Convention on biological diversity: national biodiversity strategy to 2020 and action plan to 2010. Lao People's Democratic Republic, Vientiane

NBSAP (2010) Fourth national report to the convention on biological diversity. Lao People's Democratic Republic, Vientiane

NIAID (2007) Multidrug-resistant tuberculosis (MDR TB) and possible effective treatments. National Institute of Allergy and Infectious Diseases: Understanding TB. U.S. Department of Health and Human Services—National Institutes of Health. http://www.niaid.nih.gov/topics/tuberculosis/Understanding/WhatIsTB/VisualTour/pages/mdr-tb.aspx. Accessed 18 Feb 2013

NIAID (2010) Understanding TB. National Institute of Allergy and Infectious Diseases—U.S. Department of Health and Human Services—National Institutes of Health. http://www.niaid.nih.gov/topics/tuberculosis/Understanding/pages/overview.aspx. Accessed 18 Feb 2013

Nicholas GM, Blunt JW, Munro MH (2001) Cortamidine oxide, a novel disulfide metabolite from the New Zealand basidiomycete (mushroom) *Cortinarius* species. J Nat Prod 64(3):341–344

NLL (2009) About DLLM. National library of Laos: the digital library of Lao manuscripts. http://laomanuscripts.net/en/pages/about.html. Accessed 19 Feb 2013

O'Donnell G, Poeschl R, Zimhony O, Gunaratnam M, Moreira JBC, Neidle S, Evangelopoulos D, Bhakta S, Malkinson JP, Boshoff HI, Lenaerts A, Gibbons S (2009) Bioactive Pyridine-N-oxide Disulfides from *Allium stipitatum*. J Nat Prod 72(3):360–365

Olson DM, Dinerstein E (2002) The global 200: priority ecoregions for global conservation. Ann Mo Bot Gard 89(1):199–224

Patil CL, Maripuu T, Hadley C, Sellen DW (2012) Identifying gaps in health research among refugees resettled in Canada. International Migration. DOI:10.1111/j.1468-2435.2011.00722.x

Pfeiffer JM, Butz RJ (2005) Assessing cultural and ecological variation an ethnobiological research: the importance of gender. J Ethnobiol 25(2):240–278

Phiapalath P (2009) Distribution, behavior and threat of red-shanked Douc Langur *Pygathrix nemaeus* in Hin Namno national protected area. Suranaree University of Technology, Khammouane

Pholsena V, Banomyong R (2006) Laos: from buffer state to crossroads? 2004 (French language). Trans. Smithies, M. Mekong, Chiang Mai

Pierre L (1881) Flore Forestiere de la Cochinchine, vol 2. Octove Doin, Paris

Pottier R (2007) Yu Di Mi Heng; Etre Bien. Avoir de la Force; Essai Sur Les Pratiques Therapeutiques Lao. Ecole Francaise D'extreme Orient, Paris

Reich D, Patterson N, Kircher M, Delfin F, Nandineni MR, Pugach I, Ko AM-S, Ko Y-C, Jinam TA, Phipps ME, Saitou N, Wollstein A, Kayser M, Paabo S, Stoneking M (2011) Denisova admixture and the first modern human dispersals into Southeast Asia and Oceania. Am J Hum Genet 89:516–528

Riley M (2000) The traditional medicine research center (TMRC): a potential tool for protecting traditional and tribal medicinal knowledge in Laos. Cultural Survival 24(4):21–24

Roszak T (1993) The voice of the earth. Simon & Schuster, New York

Saslis-Lagoudakis CH, Savolainen V, Williamson EM, Forest F, Wagstaff SJ, Baral SR, Watson MF, Pendry CA, Hawkins JA (2012) Phylogenies reveal predictive power of traditional medicine in bioprospecting. Proc Natl Acad Sci U S A 109(39):15835–15840

Schultes RE (1963) The widening panorama in medical botany. Rhodora 65(762):97–120

Shaw S, Cosbey A, Baumuller H, Collander T, Sylavong L (2007) Rapid trade and environment assessment (RTEA). National report for Lao PDR, Winnipeg

Soejarto DD, Gyllenhaal C, Regalado J, Pezzuto J, Fong H, Tan GT, Hiep NT, Xuan LT, Binh DQ, Van Hung N, Bich TQ, Thin NN, Loc PK, Vu BM, Southavong BH, Sydara K, Bouamanivong S, O'Neill M, Lewis J, Xie X, Dietzman G (1999) Studies on biodiversity of Vietnam and Laos: the UIC-based ICBG program. Pharm Biol 37(4):100–113

Soejarto DD, Gyllenhaal C, Regalado JC, Pezzuto JM, Fong HS, Tan GT, Hiep NT, Xuan LT, Hung N, Bich TQ, Loc PK, Vu BM, Southavong B, Sydara K, Bouamanivong S, O'Neill MJ, Dietzman G (2002) An international collaborative program to discover new drugs from tropical biodiversity of Vietnam and Laos. Nat Prod Sci 8:1–15

Soejarto DD, Gyllenhaal C, Fong HH, Xuan LT, Hiep NT, Hung NV, Bich TQ, Southavong B, Sydara K, Pezzuto JM (2004) The UIC ICBG (University of Illinois at Chicago International Cooperative Biodiversity Group) Memorandum of Agreement: a model of benefit-sharing arrangement in natural products drug discovery and development. J Nat Prod 67(2):294–299

Soejarto DD, Southavong B, Sydara K, Bouamanivong S, Riley MC, Libman AS, Kadushin MR, Gyllenhaal C (2009) Studies on medicinal plants of Laos—a collaborative program between the University of Illinois at Chicago (UIC) and the Traditional Medicine Research Center (TMRC), Lao PDR: accomplishments 1998–2005. In: Compton CJ, Hartmann JF, Sysamouth V (eds) Contemporary Lao Studies: research on development, culture, language, and traditional medicine. Center for Lao Studies and the Center for Southeast Asian Studies, San Fransisco, pp 307–323

Soejarto DD, Gyllenhaal C, Kadushin MR, Southavong B, Sydara K, Bouamanivong S, Xayveu M, Zhang HJ, Franzblau SG, Tan G, Pezzuto JM, Riley MC, Elkington BG, Waller DP (2012) An ethnobotanical survey of medicinal plants of Laos toward the discovery of bioactive compounds as potential candidates for pharmaceutical development. Pharm Biol 50(1):42–60

Sommerville M (2010) Siam on the Meinam from the gulf to Ayuthia: together with three romances illustrative of Siamese life and customs. 1897 by J.B. Lippincott Company. White Lotus, Bangkok

Stibig HJ, Stolle F, Dennis R, Feldkötter C (2007) Forest cover change in Southeast Asia—the regional pattern. European Union Institute for Environment and Sustainability, Ispra

Stuart-Fox M (1997) A history of Laos. Cambridge University, Cambridge

Sydara K (2007) Environmental impacts of trade liberalization in the medicinal plants & spices sector. IISD, Lao PDR

Sydara K, Gneunphonsavath S, Wahlstrom R, Freudenthal S, Houamboun K, Tomson G, Falkenberg T (2005) Use of traditional medicine in Lao PDR. Complement Ther Med 13(3):199–205

Takayama H (2004) Chemistry and pharmacology of analgesic indole alkaloids from the rubiaceous plant, *Mitragyna speciosa*. Chem Pharm Bull(Tokyo) 52(8):916–928

Tayles N, Buckley HR (2004) Leprosy and tuberculosis in Iron Age Southeast Asia? Am J Phys Anthropol 125(3):239–256

Thompson C, Thompson T (2008) First contact in the Greater Mekong: new species discoveries 1997–2007. World Wildlife Fund, Hanoi

Tiyavanich K (2003) The Buddha in the Jungle. Silkworm Books, Chiang Mai

Tossa W (2002) Khun Bulomrajathirat or Lord Bulomrajathirat. SEAsite Laos. http://www.seasite. niu.edu/lao/laoliterature/Lao_traditional_literature/khun_borom.htm. Accessed 19 Feb 2013

Tropicos.org (2011). Tinospora crispa. Missouri Botanical Garden. http://www.tropicos.org/ Name/50053822?tab=synonyms. Accessed 19 Feb 2013

UN (1992) Convention on biological diversity. Rio De Janeiro

UN (2011) Nagoya protocol on access and benefit-sharing. Secretariat of the Convention on Biological Diversity, Montreal

UNDP (2011) Lao People's Democratic Republic. International human development indicators. http://hdrstats.undp.org/en/countries/profiles/LAO.html. Accessed 19 Feb 2013

UNDP (2012) Lao PDR: environment and energy. http://www.undplao.org/whatwedo/energy_env. php. Accessed 19 Feb 2013

UNEP-WCMC (2011) UNEP_WCMC species database: CITES-Listed species. http://www.cites. org/eng/resources/species.html. Accessed 19 Feb 2013

Ventura M, Canchaya C, Tauch A, Chandra G, Fitzgerald GF, Chater KF, van Sinderen D (2007) Genomics of actinobacteria: tracing the evolutionary history of an ancient phylum. Microbiol Mol Biol Rev 71(3):495–548

Vidal J (1959) Noms Vernaculaires de Plantes en Usage au Laos. Extrait du Bulletin de l'Ecole Francaise d'Extrême-Orient, Tome XLIX, fascicule 2. L'Ecole Francaise d'Extrême-Orient (EFEO), Paris

Viravong MS (1964) History of Laos. Paragon Book Reprint, New York

Vongsavanh S (1984) RLG military operations and activities in the Laotian Panhandle. United States Army Center of Military History, Washington

WHO (2003) WHO guidelines on good agricultural and collection practices (GACP) for medicinal plants. World Health Organization, Geneva

WHO (2006) Tuberculosis fact sheets: what is DOTS? Communicable Diseases Department. http://www.searo.who.int/en/Section10/Section2097/Section2106_10678.htm. Accessed 19 Feb 2013

WHO (2008a) International Agency for Research on Cancer: GLOBOCAN 2008 Fast Stats: Lao People's Democratic Republic. World Health Organization. http://globocan.iarc.fr/factsheet. asp. Accessed 19 Feb 2013

WHO (2008b) Media centre: fact sheet No 134. Traditional Medicine. World Health Organization. http://www.who.int/mediacentre/factsheets/fs134/en/. Accessed 19 Feb 2013

WHO (2010a) Lao People's Democratic Republic: Tuberculosis profile. World Health Organization. https://extranet.who.int/sree/Reports?op=Replet&name=/WHO_HQ_Reports/G2/PROD/EXT/TBCountryProfile&ISO2=LA&outtype=html. Accessed 19 Jan 2012

WHO (2010b) Tuberculosis fact sheet No. 104. http://www.who.int/mediacentre/factsheets/fs104/en/index.html. Accessed 11 Dec 2011

WHO (2011). World Health Organization HIV/TB Factsheet

WHO (2012a) Lao People's Democratic Republic: health profile. http://www.who.int/gho/countries/lao.pdf. Accessed 19 Feb 2013

WHO (2012b) Media centre: tuberculosis. Fact sheet No. 104. World Health Organization, Geneva. http://www.who.int/mediacentre/factsheets/fs104/en/. Accessed 19 Feb 2013

WHO (2012c) World Health Organization: search results. http://search.who.int/search?q=fever&ie=utf8&site=default_collection&client=_en&proxystylesheet=_en&output=xml:no_dtd&oe=utf8. Accessed 19 Feb 2013

Wikipedia (2012) Laos. Wikipedia, the free encyclopedia. http://en.wikipedia.org/wiki/Laos. Accessed 19 Feb 2013

WorldBank (2006) Sector plan for sustainable development of the mining sector in Lao PDR: final report. World Bank, Vientiane

WWF (2012a) Mekong River Basin. Damming the Mekong. World Wildlife Fund. Washington. http://wwf.panda.org/what_we_do/footprint/water/dams_initiative/examples/mekong/. Accessed 19 Feb 2013

WWF (2012b) Rapid species discoveries make the case for a Greener Mekong. Focus 34(2):7

Wyatt DK (1986) Thailand: a short history. Yale University Press, New Haven

Xue B, Su YCF, Mols JB, Keßler PJA, Saunders RMK (2011) Further fragmentation of the polyphyletic genus *Polyalthia* (Annonaceae): molecular phylogenetic support for a broader delimitation of *Marsypopetalum*. Systematics and Biodiversity 9(1):17–26

Xueqin C (1974) The Dreamer Wakes. Trans. Minford J. In: Gao E (Ed.). The Story of the Stone. Vol. 5. Penguin Classics, London, reprint

Zasloff JJ, Langer PF (1970) North Vietnam and the Pathet Lao: partners in the struggle for Laos, 1st edn, Rand Corporation Research Studies. Harvard University Press, Cambridge